W9-CKH-040

Science and Religion, 1450–1900

From Copernicus to Darwin

RICHARD G. OLSON

Greenwood Guides to Science and Religion

Greenwood Press
Westport, Connecticut • London

Library of Congress Cataloging-in-Publication Data

Olson, Richard, 1940–
Science and religion, 1450–1900 : from Copernicus to Darwin / Richard G. Olson.
p. cm.—(Greenwood guides to science and religion)
Includes bibliographical references and index.
ISBN 0–313–32694–0 (alk. paper)
1. Religion and science—History. I. Title. II. Series.
BL245.O47 2004
261.5'5'09—dc22 2004047501

British Library Cataloguing in Publication Data is available.

Library of Congress Catalog Card Number: 2004047501
ISBN: 0–313–32694–0

First published in 2004

Greenwood Press, 88 Post Road West, Westport, CT 06881
An imprint of Greenwood Publishing Group, Inc.
www.greenwood.com

Printed in the United States of America

The paper used in this book complies with the Permanent Paper Standard issued by the National Information Standards Organization (Z39.48–1984).

10 9 8 7 6 5 4 3 2 1

Copyright Acknowledgments

The author and publisher gratefully acknowledge permission to quote passages from the following sources:

Robert Boyle, "A Free Inquiry into the Vulgarly Conceived Notion of Nature," in *The Works of the Honourable Robert Boyle. In Six Volumes*, Volume 5 (London: 1772). Courtesy of the Honnold/Mudd Library.

Thomas Burnet, *The Theory of the Earth: Containing an Account of the Original of the Earth, and All of the General Changes Which It Hath Already Undergon or Is to Undergo Till the Consummation of All Things*, second edition (London: 1691). Courtesy of the Honnold/Mudd Library.

John Ray, *The Wisdom of God Manifested in the Works of Creation*, third edition (London: 1705). Courtesy of the Claremont School of Theology.

Contents

Illustrations

Series Foreword

For nearly 2,500 years, some conservative members of society have expressed concern about the activities of those who sought to find a naturalistic explanation for natural phenomena. In 429 B.C.E., for example, the comic playwright, Aristophanes parodied Socrates as someone who studied the phenomena of the atmosphere, turning the awe-inspiring thunder that had seemed to express the wrath of Zeus into nothing but the farting of the clouds. Such actions, Aristophanes argued, were blasphemous and would undermine all tradition, law, and custom. Among early Christian spokespersons there were some, such as Tertullian, who also criticized those who sought to understand the natural world on the grounds that they "persist in applying their studies to a vain purpose, since they indulge their curiosity on natural objects, which they ought rather [direct] to their Creator and Governor" (1896–1903, 3:133).

In the twentieth century, though a general distrust of science persisted among some conservative groups, the most intense opposition was reserved for the theory of evolution by natural selection. Typical of extreme anti-evolution comments is the following opinion offered by Judge Braswell Dean of the Georgia Court of Appeals: "This monkey mythology of Darwin is the cause of permissiveness, promiscuity, pills, prophylactics, perversions, pregnancies, abortions, pornography, pollution, poisoning, and proliferation of crimes of all types" (Toumey 1994, 94).

It can hardly be surprising that those committed to the study of natural phenomena responded to their denigrators in kind, accusing them of willful ignorance and of repressive behavior. Thus, when Galileo Galilei was warned against holding and teaching the Copernican system of astronomy as true, he wielded his brilliantly ironic pen and threw down a gauntlet to religious authorities in an introductory letter "To the Discerning Reader" at the beginning of his great *Dialogue Concerning the Two Chief World Systems*:

> Several years Ago there was published in Rome a salutory edict which, in order to obviate the dangerous tendencies of our age, imposed a seasonable silence upon the Pythagorean [and Copernican] opinion that the earth moves. There were those who impudently asserted that this decree had its origin, not in judicious inquiry, but in passion none too well informed. Complaints were to be heard that advisors who were totally unskilled at astronomical observations ought not to clip the wings of reflective intellects by means of rash prohibitions.
>
> Upon hearing such carping insolence, my zeal could not be contained. (1953, 5)

No contemporary discerning reader could have missed Galileo's anger and disdain for those he considered enemies of free scientific inquiry.

Even more bitter than Galileo was Thomas Henry Huxley, often known as "Darwin's bulldog." In 1860, after a famous confrontation with the Anglican bishop Samuel Wilberforce, Huxley bemoaned the persecution suffered by many natural philosophers, but then he reflected that the scientists were exacting their revenge:

> Extinguished theologians lie about the cradle of every science as the strangled snakes beside that of Hercules; and history records that whenever science and orthodoxy have been fairly opposed, the latter has been forced to retire from the lists, bleeding and crushed, if not annihilated; scotched if not slain. (Moore 1979, 60)

The impression left, considering these colorful complaints from both sides, is that science and religion must continually be at war with one another. That view was reinforced by Andrew Dickson White's *A History of the Warfare of Science with Theology in Christendom*, which has seldom been out of print since it was published as a two-volume work in 1896. White's views have shaped lay understanding of science and religion interactions for more than a century, but recent and more careful scholarship has shown that confrontational stances do not represent the views of the overwhelming majority of either scientific investigators or religious figures throughout history.

One response among those who wish to deny that conflict constitutes the most frequent relationship between science and religion is to claim that such conflict can not exist because these pursuits address completely different human needs and therefore have nothing to do with one another. This was the position of Immanuel Kant who insisted that the world of natural phenomena, with its dependence on deterministic causality, is fundamentally disjoint from the noumenal world of human choice and morality, which constitutes the domain of religion. Much more recently, it was the position taken by Stephen Jay Gould in *Rocks of Ages: Science and Religion in the Fullness of Life*. Gould writes:

> I . . . do not understand why the two enterprises should experience any conflict. Science tries to document the factual character of the natural world and to develop theories that coordinate and explain these facts. Religion, on the other hand, operates in the equally important, but utterly different realm of human purposes, meanings, and values. (1999, 4)

In order to capture the disjunction between science and religion, Gould enunciates a principle of "Non-overlapping magisteria," which he identifies as "a principle of respectful noninterference" (1999, 5).

In spite of the intense desire of those who wish to isolate science and religion from one another in order to protect the autonomy of one, the other, or both, there are many reasons to believe that this is ultimately an impossible task. One of the central issues addressed by many religions is the relationship between members of the human community and the natural world. This is a central question addressed in Genesis, for example. Any attempt to relate human and natural existence depends heavily on the understanding of nature that exists within a culture. So where nature is studied through scientific methods, scientific knowledge is unavoidably incorporated into religious thought. The need to interpret Genesis in terms of the dominant understandings of nature thus gave rise to a tradition of scientifically informed commentaries on the six days of creation, which constituted a major genre of Christian literature from the early days of Christianity through the Renaissance.

It is also widely understood that in relatively simple cultures—even those of early urban centers—there is a low level of cultural specialization, so economic, religious, and knowledge-producing specialties are highly integrated. In Bronze-Age Mesopotamia, for example, agricultural activities were governed both by knowledge of the physical conditions necessary for successful farming and by religious rituals associated with plowing, planting, irrigating, and harvesting. Thus, religious practices and natural knowledge interacted to establish the character and timing of farming activities.

Even in very complex industrial societies with high levels of specialization and division of labor, the various cultural specialties are never completely isolated from one another and they share many common values and assumptions. Given the linked nature of virtually all institutions in any culture, it is the case that when either religious or scientific institutions change substantially, those changes are likely to produce pressure for change in the other. It was probably true, for example, that the attempts of pre-Socratic investigators of nature, with their emphasis on uniformities in the natural world and apparent examples of events systematically directed toward particular ends, made it difficult to sustain beliefs in the old pantheon of human-like and fundamentally capricious Olympian gods. But it is equally true that the attempts to understand nature promoted a new notion of the divine—a notion that was both monotheistic and transcendent, rather than polytheistic and immanent—that focused on both justice and intellect rather than power and passion. Thus, early Greek natural philosophy undoubtedly played a role not simply in challenging but also in transforming Greek religious sensibilities.

Transforming pressures do not always run just from scientific to religious domains, however. During the Renaissance, there was a dramatic change of thought among Christian intellectuals from one that focused on the contemplation of God's works to one that focused on the responsibility of the Christian to care for his fellow humans. The active life of service to humankind, rather than the contemplative life of reflection on God's character and works, now became the Christian ideal for many. As a consequence of this new focus on the active life, Renaissance intellectuals turned away from the then dominant Aristotelian view of science, which saw the inability of theoretical sciences to change the world as a positive virtue. They replaced this understanding with a new view of natural knowledge, promoted in the writings of men such as Johann Andreae in Germany and Francis Bacon in England, which viewed natural knowledge as significant only because it gave humankind the ability to manipulate the world to improve the quality of life. Natural knowledge would henceforth be prized by many because it conferred power over the natural world. Modern science thus took on a distinctly utilitarian shape, a response due at least in part to religious changes.

Neither the conflict model nor the claim of disjunction, then, accurately reflect the often intense and frequently supportive interactions between religious institutions, practices, ideas, and attitudes on the one hand, and scientific institutions, practices, ideas, and attitudes on the other. Without denying the existence of tensions, the primary goal of this series is to explore the vast domain of mutually supportive and/or transformative interactions between scientific institutions, practices, and knowledge and religious institutions, practices, and beliefs. A second goal is to offer the opportunity to make comparisons across space, time, and cultural configuration. The series will cover the entire globe, most major faith traditions, hunter-gatherer societies in Africa and Oceana as well as advanced industrial societies in the West, and the span of time from classical antiquity to the present. Each volume will focus on a particular cultural tradition, faith community, time period, or scientific domain, so that each reader can enter the fascinating story of interactions between science and religion from a familiar perspective. Furthermore, each volume will include not only a substantial narrative or interpretive core, but also a set of primary documents, which will allow the reader to explore relevant evidence, an extensive annotated bibliography to lead the curious to reliable scholarship on the topic, and a time line to help the reader keep track of the sequence of events involved and to relate them to major social and political occurrences.

So far I have used the words "science" and "religion" as if everyone knows and agrees about their meaning and as if they were equally appropriately applied across place and time. Neither of these assumptions is true. Science and religion are modern terms that reflect the way that we in the industrialized West organize our conceptual lives. Even in the modern West, what we mean by science and religion is likely to depend on our political orientation, our scholarly background, and the faith community to which we belong. Thus, for example, Marxists and Socialists tend to focus on the application of natural knowledge as the key element in defining science. According to the British Marxist scholar Benjamin Farrington, "Science is the system of behavior by which man has acquired mastery of his environment. It has its origins in techniques . . . in various activities by which man keeps body and soul together. Its source is experience, its aims, practical, its *only* test, that it works" (1953). Many of those who study natural knowledge in pre-industrial societies are also primarily interested in knowledge as it is used and are relatively open regarding the kind of entities posited by the developers of culturally specific natural knowledge systems or "local sciences." Thus, in his *Zapotec Science: Farming and Food in the Northern Sierra of Oaxaca*, Roberto González insists that

> Zapotec farmers . . . certainly practice science, as does any society whose members engage in subsistence activities. They hypothesize, they model problems, they experiment, they measure results, and they distribute knowledge among peers and to younger generations. But they typically proceed from markedly different premises—that is, from different conceptual bases—than their counterparts in industrialized societies. (2001, 3)

Among the "different premises" is the Zapotec scientists' presumption that unobservable spirit entities play a significant role in natural phenomena.

Those more committed to liberal pluralist society and to what anthropologists like González are inclined to identify as "cosmopolitan science" tend to focus on science as a source of objective or disinterested knowledge, disconnected from its uses. Moreover, they generally reject the positing of unobservable entities, which they characterize as "supernatural." Thus, in an *Amicus Curiae* brief filed in connection with the 1986 Supreme Court case that tested Louisiana's law requiring the teaching of creation science along with evolution, *72 Nobel Laureates, 17 State Academies of Science and Seven Other Scientific Organizations* argued that

> Science is devoted to formulating and testing naturalistic explanations for natural phenomena. It is a process for systematically collecting and recording data about the physical world, then categorizing and studying the collected data in an effort to infer the principles of nature that best explain the observed phenomena. Science is not equipped to evaluate supernatural explanations for our observations; without passing judgement on the truth or falsity of supernatural explanations, science leaves their consideration to the domain of religious faith. (72 Nobel Laureates 1986, 24)

No reference whatsoever to uses appears in this definition. And its specific unwillingness to admit speculation regarding supernatural entities into science reflects a society in which cultural specialization has proceeded much farther than in the village farming communities of southern Mexico.

In a similar way, secular anthropologists and sociologists are inclined to define the key features of religion in a very different way than members of modern Christian faith communities. Anthropologists and sociologists focus on communal rituals and practices that accompany major collective and individual events: plowing, planting, harvesting, threshing, hunting, preparation for war (or peace), birth, the achievement of manhood or womanhood, marriage (in many cultures), childbirth, and death. Moreover, they tend to see the intensification of social cohesion as the major consequence of religious practices. Many Christians, on the other hand, view the primary goal of their religion as personal salvation, viewing society at best as a supportive structure and at worst as a distraction from their own private spiritual quest.

Thus, science and religion are far from uniformly understood. Moreover, they are modern Western constructs or categories whose applicability to the temporal and spatial "other" must always be justified and must furthermore be interpreted as the means by which we organize our understanding of the actions and beliefs of people who would not have used those terms themselves. Nonetheless it does seem to us not simply permissible, but probably necessary to use these categories at the start of any attempt to understand how actors from other times and places interacted with the natural world and with their fellow humans. It may ultimately be possible for historians and anthropologists to understand the practices of persons distant in time and/or space in terms that those persons might use. But that process must begin by likening the actions of others to those that we understand from our own experience, even if the likenesses are inexact and in need of qualification.

The editors of this series have not imposed any particular definition of science or of religion on the authors, expecting that each author will develop either explicit or implicit definitions that are appropriate to their own scholarly approaches and to the topics that they have been assigned to cover.

Richard Olson
Claremont, California

Acknowledgments

My special thanks go to the Liberal Arts and International Studies Division of the Colorado School of Mines. A year spent there as the Hennebach Visiting Professor allowed me time to complete a first draft of this work and offered me the opportunity to gain from conversations with Carl Mitcham, who has a long-standing interest in religion and technology. I also owe my Greenwood Press editor, Kevin Downing, a major debt for giving me the opportunity to act as editor for this series, for the chance to write this volume, and for persistently trying—occasionally successfully—to keep my mind on the audience for which I was supposed to be writing. Kathleen Mulhern, a history graduate student at the University of Colorado, Boulder, and a former editor of religious works, read the entire first draft and offered many useful comments. Except for the specific figures designated below, all of the images in the volume were produced by the special collections department of the Honnold-Mudd Library of the Claremont Colleges and appear through their permission. Figure 4.2 appears by permission of the director of the library of the Claremont School of Theology. Figures 1.2, 1.3, and 2.2 are taken from Richard Olson, *Science Deified and Science Defied, Volume 1* (1982), for which I own the copyright, and Figure 7.6 is from the first, 1844 edition of *Vestiges of the Natural History of Creation*.

Chronology of Events

1347	Onset of the Black Death in Europe.
1431	Death of Joan of Arc.
1437	Cosimo de' Medici comes to power in Florence.
1463	Marsilio Ficino begins to translate the *Hermetic Corpus.*
1492	Columbus lands in the New World.
1517	Martin Luther posts his ninety-five theses.
1528	Paracelsus leaves Basel, begins his itinerant years.
1534	The Act of Supremacy declares Henry VIII the supreme head of the English Church.
1536	Publication of John Calvin's *Institutes of the Christian Religion.*
1543	Copernicus publishes *On the Revolutions of the Heavenly Spheres.*
1545–1563	The Council of Trent initiates the Counter-Reformation.
1558	Elizabeth I ascends to the English throne.
1588	The Spanish Armada defeated by the British.
1592	Richard Hooker publishes the first four books of *The Laws of Ecclesiastical Polity.*

1598	Henry IV of France grants limited local rights to Protestants in the Edict of Nantes.
1603	James I comes to the throne in England.
1603	Shakespeare's *Hamlet* first performed.
1616	Galileo warned by Cardinal Bellarmine not to present the Copernican theory as true.
1618	Beginning of the Thirty Years' War in the Germanies.
1619	Johann Andreae's *Christianopolis.*
1626	Francis Bacon's *New Atlantis.*
1632	Galileo's *Dialogue Concerning the Two Chief World Systems.*
1633	Galileo found guilty of grave suspicion of heresy.
1640	The "Long Parliament" begins the English Civil War, which lasts until 1660.
1644	René Descartes publishes his *Principles of Natural Philosophy.*
1651	Thomas Hobbes' *Leviathan.*
1658	Pierre Gassendi's *Syntagma Philosophicum* offers a Christianized version of atomism.
1660	Restoration of the English Monarchy.
1660	Robert Boyle's *New Experiments Physico-Mechanical. . . .*
1665	Athanasius Kircher's *Mundus Subterraneus.*
1680–1689	Thomas Burnet's *Sacred Theory of the Earth.*
1685	Louis XIV of France revokes the Edict of Nantes.
1687	Isaac Newton's *Mathematical Principles of Natural Philosophy.*
1688	The Glorious Revolution assures Protestant succession in England.
1692	First edition of John Ray's *The Wisdom of God Manifested in the Works of the Creation.*

1693 Richard Bentley uses Newtonian Science in his *Confutation of Atheism.*

1696 William Whiston's *New Theory of the Earth.*

1696 John Toland's deistic *Christianity Not Mysterious* appears.

1724 Publication of John Hutchinson's *Moses' Principia.*

1740–1748 War of the Austrian Succession.

1749–1778 Comte de Buffon's *Natural History* appears. Early volumes support uniformitarian geological understanding; later volumes offer empirical support for extension of age of the universe.

1755 Immanuel Kant's *General Natural History and Theory of the Heavens.*

1762 David Hume writes but does not publish *Dialogues Concerning Natural Religion,* does publish *The Natural History of Religion.*

1764 Thomas Reid's *An Inquiry into the Human Mind on the Principles of Common Sense.*

1776 The American Revolution begins.

1776 Adam Smith's *Wealth of Nations* appears.

1784–1791 Johann Gottfried Herder's *Reflections on the Philosophy of the History of Mankind.*

1789 The French Revolution begins.

1791 Kant's *Critique of Pure Reason* initiates his "critical philosophy."

1797 Napolean comes to power in France.

1799 Friedrich Schleiermacher's *Speeches on Religion to its Cultured Despisers.*

1803 F.W.J. Schelling's *Ideas for a Philosophy of Nature* introduces his Absolute.

1809 J. B. Lamarck's *Philosophie zoologique* offers a progressive evolutionary theory.

1811 Georges Cuvier's *Discours sur les revolutions de la surface du globe* introduces "catastrophist" geology.

1812 Napoleon defeated at Moscow.

1815–1848 Prince Klemmens von Metternich leads the conservative German Confederation.

1826 Auguste Comte begins lectures on *The Positive Philosophy.*

1830–1833 Charles Lyell's *Principles of Geology* revives uniformitarian approach to geology.

1836 William Buckland's *Geology and Minerology Considered with Reference to Natural Theology* promotes catastrophist geology as a support for Scripture.

1837–1901 Queen Victoria's reign in England.

1841 Ludwig Feuerbach's *The Essence of Christianity.*

1844 Robert Chambers' *Vestiges of the Natural History of Creation* popularizes evolutionary ideas and generates opposition.

1848 Revolutions break out throughout continental Europe beginning in Paris. They are soon quelled.

1851 Comte's work *The System of Positive Polity* proposes a "Religion of Humanity."

1859 Charles Darwin's *On the Origin of Species.*

1860 T. H. Huxley and Bishop Samuel Wiberforce confront one another over human ancestry at the Oxford meeting of the British Association for the Advancement of Science meeting.

1868 Lord Kelvin estimates the age of the earth at five million years, too short for Darwinian theory to adequately account for speciation.

1868 Ernst Haeckel's *The History of Creation: or the Development of the Earth and its Inhabitants by the Action of Natural Causes*—the most widely read account of evolution by natural selection in the nineteenth century.

1870	The Franco-Prussian War.
1871	Darwin's *The Descent of Man* argues for the evolutionary significance of religion.
1874	William Draper introduces "conflict" language in *A History of the Conflict between Religion and Science.*
1884	Charles Hodge's *What Is Darwinism?* offers an evangelical critique of Darwin's work.
1899	Haeckel's *Riddle of the Universe at the Close of the Nineteenth Century.*

Chapter 1

Introduction: Galileo and the Church—Or, How Do Science and Religion Interact?

This book will tell the story of how science and religion have interacted in the Western World from the beginning of the Scientific Revolution (about 1450) to the end of the nineteenth century. In order to focus on the historical events and characters, I will rarely discuss the often acrimonious disagreements historians have had regarding almost every topic I will be discussing. This chapter, however, will introduce readers to the many ways in which scientists, clergy, and secular scholars of the last 250 years have tried to understand the connections between science and religion. I will illustrate my own approach and contrast it with others by discussing the most famous case in the history of science and its interactions with religion—that of Galileo and members of the Catholic Church between 1615 and 1633.

One's approach to science and religion interactions will depend on how one defines the central purposes and the appropriate boundaries of science and/or religion. The philosopher Immanuel Kant (1724–1804) and the extremely influential Protestant theologian Rudolph Bultmann (1884–1976), for example, insisted there could be no authentic interactions between the two because religion refers to personal, subjective relations—what many call "I-thou" relations—while science refers to impersonal, objective "I-it" relations. For supporters of Kant and Bultmann, the Book of Genesis is not about the origins of the physical universe at all. It is an allegory whose function is to promote an appropriate understanding of and give personal meaning to a believer's earthly existence. From this perspective, any claimed interaction between science and religion involves a fundamental misunderstanding. At the opposite end of the theological spectrum the process theologians, an important group of modern liberal Protestant theologians who are followers of Alfred North Whitehead's philosophical views, insist that God is part of the natural world and co-evolves with it; so any change in scientific knowledge of nature constitutes a change in our understanding of God.

I will be taking a middle ground in this volume, but I acknowledge that Bultmann's theology is more characteristic of the positions held within the Christian tradition in the West from the beginnings of the Renaissance to the present. So, when Galileo Galilei borrowed the remark of Cardinal Cesare Baronius, claiming that the Bible tells us how to go to heaven, not how the heavens go, he was taking a position that most Christian theologians over the past 400 years would have accepted.

There are several reasons that the boundaries between science and religion are less clear and more blurred and permeable than Kant, Bultmann, and Galileo might have wished. The first is that science and religion must both use language—not just the jargon of each discipline, but also the ordinary language of common people. The meanings of the statements of scientists and theologians alike must thus depend on the common experiences and the often unrecognized shared cultural assumptions of the people who speak the language through which they approach their subjects. There is, therefore, bound to be a crossover—whether intended or not—between the concepts associated with science, or natural knowledge, and those associated with religious beliefs at any particular time and place.

This insight was to a large degree already embodied in a claim initiated by the early Alexandrine theologian Origen (186–254) that was pervasive in the early modern period. God, it was thought, expressed his nature through two "books," that of nature and that of scripture. As products of the same divine mind, the two books could be expected to provide the same or similar challenges to their human interpreters. According to Origen, "He who has once accepted the scriptures as the work of Him who created the world must be convinced that whatever difficulties in regard to creation confront those who strive to understand its system will occur also in regard to the scriptures" (Bigg 1968, 150). Modern historians have explored how science, religion, history, law, and other fields used common language, concepts, and methods of investigation in early modern England because all developed within a common cultural setting (Shapiro 1983). In particular, they have followed changes in linguistic usage and have shown how this affected both Biblical hermeneutics—the interpretation of God's word in Scripture—and the growth of modern approaches to science—the interpretation of God's other book (Bono 1995; Harrison 2001).

A second reason for porous boundaries between science and religion is that, for much of the time that we will be considering, the same individuals were often both scientists and serious religious thinkers. This was true of the early chemist Robert Boyle (1627–1691) and of the most famous of all early modern natural philosophers, Sir Isaac Newton (1642–1727), for example. Such men found it virtually impossible not to import concepts, presuppositions, and methods from one domain into the other, sometimes with important consequences. The union of scientific and religious perspectives within individual thinking did not end in the seventeenth century. Religious and scientific ideas were interpenetrated in the work of the Unitarian minister and pneumatic chemist Joseph Priestly (1733–1804) in the late eighteenth century, and in the ideas of the Sandemanean experimental physicist Michael Faraday (1791–1867) in the early nineteenth century. The approach to thermodynamics taken by such late-nineteenth-century Scottish evangelical physicists as James Clerk Maxwell (1831–1879), William Thomson (1824–1907), and Peter Guthrie Tait (1831–1901) was closely linked to their religious positions (Smith 1999). It has never been questioned that their religious commitments influenced the geological theories of such clergyman-geologists as William Buckland (1784–1856) and Adam Sedgewick (1785–1873) in the nineteenth century, and, even in the twentieth century, the insistence by Albert Einstein (1879–1955) that God does not play dice shaped his rejection of the dominant probabilistic interpretation of quantum mechanics.

A third way in which science and religion come to interact with one another is institutional rather than conceptual or biographical. Like all institutions, both religion and science incorporate practices that may involve imported knowledge. Almost all religions involve an annual cycle of rituals and celebrations whose timing depends upon a calendar coordinated with the agricultural cycle and solar year. Christian tradition, for example, sets the date of Easter in relation to the spring equinox; so in order to plan ahead for the Easter celebration and those events that must precede it, Christian churches must have recourse to astronomical knowledge that can predict the time of the equinox. This borrowing does not just go one way, moreover. For example, early experimental science incorporated the religious practice of using witnesses to establish the public character of scientific knowledge (Shapin and Shaffer 1985), which was in opposition to earlier secret traditions of knowledge about nature.

Like all institutions, science and religion also compete with one another and with other institutions for personnel, resources, and cultural authority and prestige. At times these institutional interactions can lead to mutually supportive activities. In early modern Europe, Jesuit educational institutions drew in Protestant students by offering courses in mathematics, astronomy, and natural philosophy, which those students wanted because they constituted a valuable preparation for desired military careers. As part of the curriculum in their colleges, however, the Jesuits required students to take courses in Catholic theology. In this way, both science and Catholicism gained potential converts.

More often, the competition for power and authority is expressed in ways that tend to advance either science or religion relative to the other. Since the High Middle Ages (c. 1250–1350), for example, religious and scientific factions have struggled with one another for influence in educational institutions, with scientific factions generally increasing their influence over time in the West. It was principally the Christian acceptance of the growing presence of science in European universities, compared to the successful religious resistance to the incursion of secular learning in Islamic educational institutions, that accounts for the fact that the scientific revolution occurred in Christian Europe, rather than in Islamic society (Huff 1995).

THE CONFLICT MODEL

History, like fiction, seems to demand conflict to make it interesting. Indeed, some historians and social theorists argue that there can be no historical change without conflict (Latour 1988). This is one of the reasons that many authors have claimed that there is some kind of intrinsic conflict between science and religion. A second reason is closely related to the fact that patriotic spirit in a country almost always rises when there is a credible external threat of some kind. Members of groups can almost always be counted on to pull together and to rally around one another when they feel threatened, and identifying an appropriate scapegoat relieves them of the burden of thinking through a complex set of circumstances. Thus, scientists and members of religious organizations alike often seek to build solidarity within their own communities by painting members of the other group as hostile.

A third and very important source of support for the conflict notion was first fully articulated in Auguste Comte's *Cours de Philosophie positive,* lectures delivered in Paris starting in 1826. Comte (1798–1857) articulated and defended the following "Law of Three Stages":

> Each of our leading conceptions—each branch of our knowledge—passes successively through three different theoretical conditions: the Theological, or fictitious; the Metaphysical, or abstract; and the Scientific, or positive. In other words, the human mind, by its nature, employs in its progress three methods of philosophizing, the characters of which are essentially different, and even radically opposed.: viz, the theological method, the metaphysical, and the positive. (1974, 25)

According to this law, the terms theological and positive refer to two different and mutually incompatible strategies for explaining the same phenomena, so the systems of concepts and propositions that they produce—that is, religion and science—must be incompatible as well. For one to guide the lives of people, the other must be abandoned or destroyed.

The views expressed by Comte entered Anglo-American intellectual life initially through the writings of John Stuart Mill and Henry Lewes and through the brilliant translation by Harriet Martineau, which appeared as *The Positive Philosophy* in 1855. It was this view that informed T.H. Huxley's incendiary denouncement of religion in the name of science in 1860 and Andrew Dickson White's 1896 *A History of the Warfare of Science with Theology in Christendom.* It was also this view that later informed Charles Coulston Gillispie's claim in 1960 that what early nineteenth-century geology "needed to become a science was to retrieve its soul from the grasp of theology" (1960, 299); and this view continued to inform the 1998 claims of the distinguished historian Bruce Mazlish when he wrote: "Comte was right . . . natural science has freed itself from religious modes of thought and triumphantly (at least in the realm of theory) advanced its own"; furthermore, "a scientific community suitable to the development of the human sciences . . . is hampered from coming into being by religious belief" (1998, 211). In fact, I would argue that Comte's view of the fundamental relationship between science and religion has dominated the lay understanding of their interactions throughout the past century and a half.

There are, however, strong empirical and theoretical reasons to believe that Comte's views were and are seriously misguided and misleading. The empirical reasons will become apparent as evidence is presented in subsequent chapters of this book, but some of the theoretical reasons demand consideration here. In the first place, as I have already pointed out, while there are sometimes overlapping interests, the core topics of concern to scientists and theologians are generally not the same. The concerns of the former are focused on developing general accounts of the workings of a natural world presumed to function without regard to the interests of the humans investigating it or to the interests of any super-sensible being or beings who might interact with the world. Scientists do not necessarily deny the possibility of divine interactions with the world; they simply deny that such interactions can be dealt with within the domain of science.

Nor is it the case that where there is an overlap of concerns—for example in connection with the issue of the origin of the Universe—a satisfactory scientific account obviates the possibility of or need for a theological account. As many Christian theologians and scientists argued during the seventeenth century, God may well have established natural laws as the mechanisms by which his ordinary providence is carried out. Furthermore, theological accounts address a different kind of meaning than scientific ones, focusing on the significance of events in terms of the relationships between the human and the divine. For this reason, scientific and religious accounts of the same phenomenon have been called "complementary" rather than incompatible by a number of scholars.

As a fundamentally descriptive and interpretive enterprise, science is not intended to be prescriptive—that is, it does not make claims about what ought to be rather than what is. On the other hand, the focus of religion and of much theology is prescriptive—it tries to articulate how humans ought to behave rather than how they do behave. Neo-classical economic theory, for example, claims that most human economic behavior can be understood in terms of the rationally calculated self-interest of human beings, whereas Christianity, like most contemporary major religions, argues that humans ought to be more concerned about the well-being of their fellow humans. There is no necessary contradiction between Christianity and neo-classical economics just because the latter describes a behavior that is not prescribed by the former. In fact, the Christian doctrine of the Fall suggests why the behavior of most humans does not conform to religious prescription. It is a fundamental error, then, to imagine that scientific knowledge is intended to serve the central aims of theological discourse, nor is it ordinarily the case that Christian theology addresses the issues dealt with by modern science. From this point of view, the tradition of Comtean positivism in seeing a necessary conflict between theological and scientific discourses is simply wrong.

THE CASE OF THE GALILEO AFFAIR

The fact that science and religion have different fundamental aims and would not thus seem to come into direct competition with one another suggests the possibility that what some people have viewed as conflict between science and religion might indeed involve legitimate struggle, whose features and meaning have been misinterpreted or misrepresented. In the natural and historical worlds, conflicts are almost always between entities of the same kind: between two males of the same species for possession of a female; two politicians for political power; two tribes or nations for territory; two religious organizations for converts; or between two scientists or groups of scientists submitting different theories for the approval of the scientific community.

Of course it is possible that two different kinds of things may come into conflict. But the fact that most conflict is between entities of the same kind suggests that when there is a claim of discord between science and religion, it might be wise to look carefully to see what other kinds of struggle might be going on instead. In order to illustrate this point, I would like to consider the series of events seen by conflict theorists as one of the most famous and clear cases of religion and science clashing. In 1633 Galileo Galilei (1564–1642) was brought to trial by the Holy Office of the Inquisition for vehement suspicion of heresy in connection with his support of the Copernican, sun-centered system of astronomy in *Dialogue Concerning the Two Chief World Systems* (1632) (see Figure 1.1).

As background it is worth noting that Copernicus (1473–1543) was a Catholic Church official in Krakow, Poland, and that his major work *De Revolutionibus* (1543), or *On the Revolutions of the Heavenly Spheres*, had been written in part in response to the problem of finding an astronomical theory that could accurately predict the date of Easter. It

DIALOGO
DI
GALILEO GALILEI LINCEO
MATEMATICO SOPRAORDINARIO
DELLO STVDIO DI PISA.
E Filoſofo, e Matematico primario del
SERENISSIMO
GR.DVCA DI TOSCANA.
Doue ne i congreſſi di quattro giornate ſi diſcorre
ſopra i due
MASSIMI SISTEMI DEL MONDO
TOLEMAICO, E COPERNICANO;
*Proponendo indeterminatamente le ragioni Filoſofiche, e Naturali
tanto per l'vna, quanto per l'altra parte.*

CON PRI  VILEGI.

IN FIORENZA, Per Gio:Batiſta Landini MDCXXXII.

CON LICENZA DE' SVPERIORI.

Figure 1.1. Title page from Galileo's *Dialogue Concerning the Two Chief World Systems.* Note that the printing privilege is granted from Florence rather than from Rome. Courtesy of Special Collections, Honnold/Mudd Library.

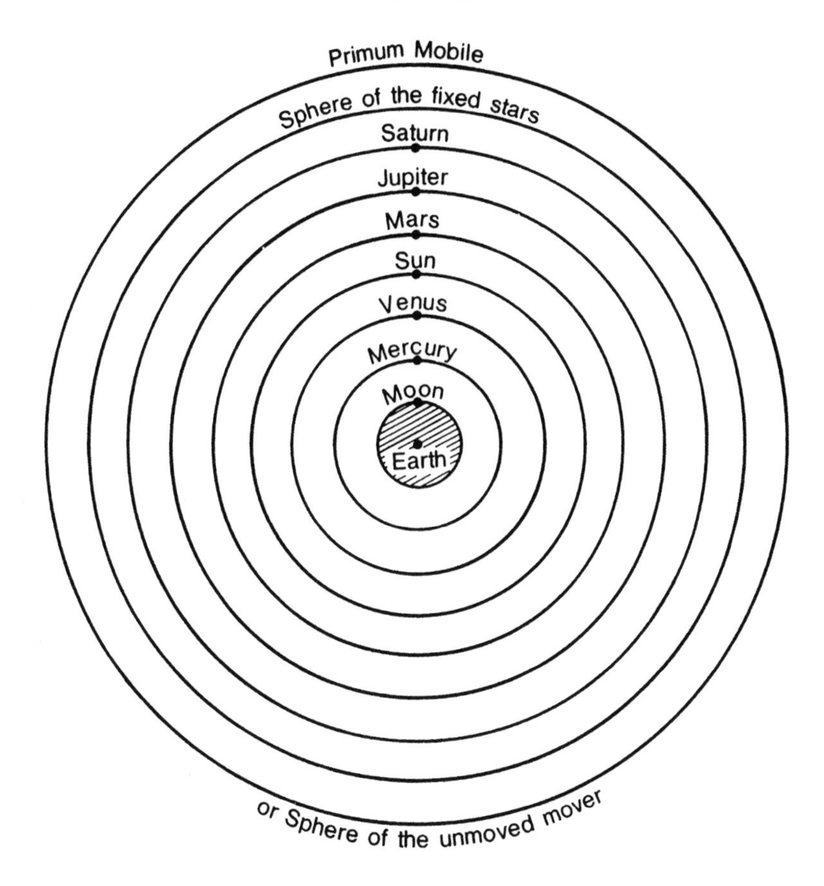

Figure 1.2. Diagram of the simplified Aristotelian Cosmos with its Earth-centered concentric spheres.

replaced the traditional understanding of the universe from antiquity, in which the earth was at the center of the universe (see Figure 1.2), with one in which the earth and all other planets revolve around the sun (see Figure 1.3).

The book appeared without a negative church response, and when the Catholic Church instituted the Gregorian calendar reform in 1586, the new calendar was based on astronomical tables calculated using the Copernican astronomical system, implying at least a limited church acceptance of Copernican astronomy. Not until 1615—nearly

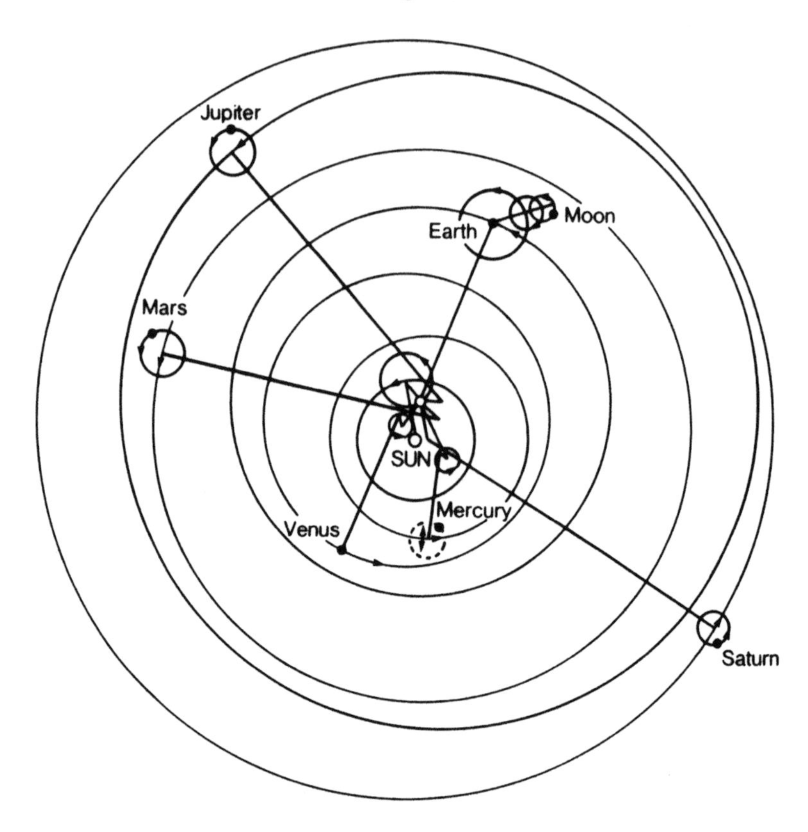

Figure 1.3. Diagram of the motion of the planets and the moon in the plane of the ecliptic in the Copernican sun-centered system.

three quarters of a century after its publication—did any official body associated with the Catholic Church take any action to suppress the Copernican theory. And when *On the Revolutions* was placed on the index of prohibited books, it was in connection with a complaint brought against Galileo by the Dominican priest Giovanni Caccini. One might wonder what triggered the actions of 1615 and 1633 if there was already some intrinsic incompatibility between Copernican astronomy and the Catholic religion as early as 1543.

In order to answer this question, we need to know something about the institutional practices and epistemological assumptions guiding astronomy and natural philosophy at the time of Copernicus and Galileo, about sixteenth- and early-seventeenth-century religious politics, and about the relationships between Galileo and several churchmen—especially between Galileo and Maffeo Barberini (1568–1644), who became Pope Urban VIII in 1623, and between Galileo and a group of Jesuit astronomers in Rome. Some authors have argued that there are key issues beyond those that I will discuss in shaping the relationship between Galileo and the Catholic Church (Redondi 1987). Most scholars would, however, probably agree that those that I point out here are the most important.

First, from late antiquity there had been a general agreement about a distinction between astronomy and natural philosophy that had become institutionalized within the medieval university and was almost universally accepted during the Renaissance. The function of astronomy was to develop mathematical hypotheses capable of saving the appearances,—that is, of calculating the observed positions of the heavenly bodies at any given time—without regard to physical reality or even to the physical possibility that a system could operate according to these hypotheses. It was the function of natural philosophers, on the other hand, to theorize about the causes of phenomena, including the phenomena of the heavens. It is now widely held that Copernicus did believe in the physical reality of a system in which the planets, including the earth, revolved about the central sun; but *On the Revolutions* contained an explicit denial, written by Copernicus's friend Andreus Osiander after Copernicus's death but before the work was published, that it was intended as anything other than a mathematical calculating scheme. As a consequence, when it appeared, the Copernican work was in conformity to the academic norms of the time, which was good because Copernicus could offer no credible explanation for the causes of the motions that his system posited.

Beginning in the thirteenth century, and growing in importance through the fourteenth and fifteenth centuries, there had been a conflict regarding the grounds of natural philosophical knowledge and the extent to which natural philosophy could offer certainty regarding the causes of phenomena. By the beginning of the sixteenth century, though there were still some who believed that Aristotelian natural philosophy could produce certain knowledge, a new philosophy associated with the term Nominalism was coming to dominate both conservative and avant-garde intellectual life. According to this philosophical perspective, human experience, which forms the foundation for scientific knowledge, is incapable of yielding certain knowledge about the causes of things. The best one can do is what astronomers did for the motion of the heavenly bodies—devise a hypothetical scheme both consistent with the available evidence and internally consistent.

Such a philosophy had significant appeal to conservative theologians because it allowed for God's unrestricted ability to arrange the world as he chose and did not seem to bind him into following any particular course of action. It also had great appeal to more adventuresome intellectuals because it freed them to speculate more creatively about the possible causes of phenomena while avoiding the nasty problem of how to guarantee the universality of generalizations induced from a limited number of examples. No matter how many times we observe the sun rising in the east, for example, there seems no way to guarantee that it will rise in the same place tomorrow. We may all agree that anyone who does not expect the sun to rise in the east is crazy; but that does not make its next appearance necessary.

For most purposes, Galileo accepted this nominalist theory of knowledge; but at times he and some of his students seemed to argue that they could prove the physical reality of the Copernican system. In doing so they came into conflict with the dominant theories of scientific method, which were also favored by many churchmen. It is hard to see criticisms of Galileo's science grounded in nominalist principles—even when articulated by Catholic Church officials—as based on a conflict between science and religion. It is, however, certainly true that Catholic Church officials used nominalist arguments to the benefit of church ideology.

By the mid-sixteenth century, the Protestant Reformation, which had been spearheaded by Martin Luther (1483–1546) in the 1490s, was well underway, and the Catholic Church was organizing its Counter-Reformation response at the Council of Trent. One of the most important issues faced at Trent involved what flexibility individuals were to be given in interpreting the meaning of the Scriptures. In theory, Luther, John Calvin (1509–1564), and most Protestants insisted upon the right of each individual Christian to determine for him- or herself the meaning of the Bible. The Catholic Church, on the other hand, affirmed at Trent a long-standing claim that only the church fathers and church councils had the authority and ability to determine the meaning of Scripture in connection with central issues of faith and morals. On April 8, 1546, it issued a decree, a portion of which stated the following:

> no one, relying on his own judgment shall, in matters of faith and morals pertaining to the edification of Christian doctrine, distorting the Holy Scriptures

in accordance with his own conceptions, presume to interpret them contrary to that sense which Holy Mother Church, to whom it belongs to judge of their true sense and interpretation, has held and holds, or even contrary to the unanimous teachings of the Fathers, even though such interpretations shall never at any time be published. (Finocchiaro 1989, 12)

The restriction here to matters of faith and morals is important, for it seemed to many good Catholics, including Galileo, to leave open the right to offer new interpretations of Scripture regarding matters not directly related to faith and morals. Such an interpretation of the Trentine decree was consistent with a long-standing doctrine first articulated by Clement of Alexandria (c. 150–215) and subsequently affirmed by no less an authority than Saint Augustine (354–430).

According to this doctrine of accommodation, a distinction must be made between Biblical teachings on spiritual matters and its descriptions of the natural world. Because a treatment of the physical world was incidental to the spiritual purposes of his revealed word, God chose to accommodate his language to the understandings of the common people in order to increase the acceptability of his spiritual message. Indeed, both Clement and Augustine had insisted that in many cases the language of the Bible in respect to physical aspects of the world could not be literally true. Because people's understandings of the natural world changed over time, it was quite allowable and appropriate to reinterpret the meaning of those scriptural passages that described or discussed the natural world. On the other hand, spiritual issues and issues connected with moral injunctions were not subject to reinterpretation based on changing natural knowledge.

Many members of the Dominican order, which was charged with defense of the faith, took a more conservative and less traditional view, holding—like later Protestant fundamentalists—that there was a single, timeless, correct interpretation of every biblical statement and that in all cases where it was possible, that interpretation should be literal. There was, then, a conflict within the Catholic Church in connection with the relationship of Scripture to knowledge of the natural world. Many church fathers, including Augustine, effectively allowed that interpretation of Scripture relative to the natural world may be controlled by the best natural knowledge available. To liberal Catholics at the beginning of the seventeenth century, this proposition seemed consistent with the decrees of the Council of Trent.

Sometime around 1613 in his conversations and correspondence with the Benedictine priest Bennedetto Castelli, and then in a famous

1615 letter to the Tuscan Grand Duchess Christina, Galileo explicitly argued that Scripture could and should be interpreted as consistent with the Copernican theory rather than with an earth-centered cosmology (Finocchiaro 1989, 114–118). Almost immediately Father Caccini, a Dominican priest, complained to the Inquisition; and on March 20, 1615, he testified to the Holy Office in Rome that he had warned Galileo's students "that no one was allowed to interpret divine Scripture in a way contrary to the sense in which all the Holy Fathers agree since this was prohibited by the Council of Trent" (Finocchiaro 1989, 137). Thus, Galileo stepped into an ongoing conflict among churchmen regarding the interpretation of Scripture.

Galileo was absolved of any wrongdoing in 1616, suggesting that the official position of the Church was consistent with Galileo's liberal interpretation of Trentine doctrine. He was, however, warned by Cardinal Bellarmine (1542–1621) on behalf of the Inquisition that he should not hold that the Copernican theory was physically true, both because such a position was inconsistent with the legitimate claims of natural philosophy according to the best nominalist assumptions and because to make such a claim in the current environment might undermine the faith of those who were not philosophically sophisticated. Furthermore, on March 5, 1616, by a decree of the index, Copernicus's *De Revolutionibus* was suspended pending correction "in order that this opinion may not creep any further to the prejudice of Catholic truth" (Finocchiaro 1989, 149). But within four years, a series of minor corrections of passages that seemed to suggest the physical reality of the earth's motion had been made and the suspension was lifted. Members of the Jesuit Order continued to teach Copernican astronomical theory as a hypothesis in their colleges throughout the seventeenth and eighteenth centuries.

There are two critically different written versions of what Bellarmine told Galileo. One, dated May 26, 1616, signed by Bellarmine and given to Galileo as a record of what was told to him, stated simply that the doctrine of Copernicus "cannot be defended or held" (Finocchiaro, 153). The other, whose origin is of debated authenticity, is in the Inquisition minutes (De Santillana 1955, 125–131). It is dated February 25, 1616, has no signature, and states that Galileo is to "abstain completely from teaching or defending [the position that the sun stands still at the center of the world and the earth moves] *or from discussing it*" (Finocchiaro 1989, 147; emphasis added). This second version of what was said is much more restrictive; the first, according to common usage of the time, allowed Galileo to continue to advocate the consideration of Copernican theory as a hypothesis, while the second demanded absolute silence.

At this point it is worth asking whether what happened to Galileo so far represents a conflict between science and religion, or what Andrew Dickson White called a "War Upon Galileo." It is clear that church officials were—probably correctly—concerned that the Copernican theory, which seemed to violate the traditional sense of the Bible, constituted a threat to the belief of ordinary lay Catholics during a period when Catholic orthodoxy was under siege. It is also clear that, given the prevailing standards in natural philosophy, Bellarmine, among others, felt justified in asking Galileo not to insist upon the physical truth of the Copernican theory. This did not mean that officials at the highest level wanted to suppress discussion of the theory as a hypothesis. White quite incorrectly characterized the position of Bellarmine and many of his fellows as follows:

> By far the most terrible champion who now appeared was Cardinal Bellarmin [*sic*], one of the greatest theologians the world has known. He was earnest, sincere, and learned, but insisted on making science conform to Scripture. The weapons which men of Bellarmin's [*sic*] stamp used were purely theological. They held up before the world the dreadful consequences which must result to Christian Theology were the heavenly bodies proved to revolve about the sun and not about the earth. (1965, 129–130)

Whether these men would have tried to suppress the Copernican opinion without their intellectually viable nominalist philosophical position, we simply do not and cannot know. What is certain about Bellarmine, however, is that none of his arguments regarding the Copernican theory were, strictly speaking, theological. They were all either about proper scientific methods or purely pragmatic. Furthermore, Bellarmine explicitly agreed with Galileo that if at some future time someone could prove the reality of the Copernican position, then a reinterpretation of Scripture would have to be undertaken (Blackwell 1991).

When White described the Galileo affair, he also made a claim that Galileo was careful never to make explicit in public—that "[Galileo's] discoveries had clearly taken the Copernican theory out of the list of hypotheses, and had placed it before the world *as a truth*" (White 1965, 126; emphasis mine). I am virtually certain that this is what Galileo believed, in spite of his careful, overt denials. But if that is so, the intellectual elite churchmen had viable scientific reasons to oppose him, given the standards of the scientific community of the time. Though Galileo's telescopic discoveries did undermine the older, Ptolemaic, earth-centered astronomy, they did not provide clear evidence for the Copernican theory. In fact, Tycho Brahe (1546–1601), one of the most distinguished astronomers of the time, developed an alternative to both the Ptolemaic and the Copernican theories that was consistent with Galileo's observations. Galileo did believe that he had evidence from the motion of the tides that the Copernican theory was correct; but by both seventeenth- and twenty-first-century scientific standards, Galileo's tidal theory was incorrect. It was not until James Bradley's discovery of the aberration of light in 1729 that there was credible direct evidence for the revolution of the earth around the sun. Thus, it is very hard to see evidence up to 1616 of religion—embodied in the official offices of the Catholic Church—ruthlessly repressing scientific speculation.

The position taken by men like Father Caccini was far more repressive, but it was not the position of any official church body, much less a position that anyone would argue was essential to religion in general. It was the position held by one segment of one religious community. White was correct in arguing that men like Caccini wanted to make science conform to their understanding of Scripture and in suggesting that this position was likely to lead to a conflict with scientists—but that hardly constitutes a war led by the Catholic Church against science. At other times, the shoe was on the other foot. In Victorian England, for example, a group of scientists, including T. H. Huxley (1824–1895) and John Tyndall (1820–1893), became identified with a doctrine called "scientific naturalism," which insisted that the only valid way of obtaining any kind of authentic knowledge was empirical. Thus, they sought to limit the ways in which religious knowledge could be discovered. In this case, theologians understandably resisted restrictions on their activities and responded by criticizing science for its repressive stance—though in this case it is also very likely that those taken to represent science were in a relatively small minority.

Before we go on to the events leading up to Galileo's trial in 1632, we need to explore a bit more about Galileo's character and his personal relations with some key players. Galileo was by all accounts arrogant and inclined to give himself credit for inventions and discoveries to which others had a claim. Consequently, he was often involved in priority disputes and litigation. For present purposes, it is significant that he became involved in a priority dispute regarding the discovery of sunspots with the Jesuit astronomer Christopher Scheiner, whose observations of the spots were discussed in a letter from a German businessman and intellectual, Mark Welser, to Galileo before Galileo published his own *Sunspot Letters* claiming sole credit for their discovery. This dispute alienated Galileo from a group who might otherwise have been willing to offer him support—up until 1613 the Jesuit astronomers in Rome had been among Galileo's strongest advocates.

When in Rome during 1611, Galileo had met and befriended Cardinal Maffeo Barberini, a student of natural philosophy who had been put in charge of the water supply to the Vatican. The two continued their friendship, which was grounded in common interests in science and literature, and Barberini dedicated a book of poetry to Galileo in 1620. Then, in 1623, Barberini was elected to the papacy under the name Urban VIII. During the next year, Galileo had a series of six meetings with the pope to discuss plans for a dialogue that would consider the relative virtues of the Ptolemaic and Copernican systems. Though there is no written record of these conversations, there is evidence from subsequent correspondence that the pope suggested the title—*Dialogue Concerning the Two Chief World Systems*—that the work was eventually published under, that he was concerned only that Galileo agree to keep the discussion at the level of hypotheses, and that he wanted Galileo expressly to acknowledge the authority of the Church in theological matters. All of these conditions seem to have been accepted by Galileo, who wrote to his friend Prince Cesi that the pope had assured him that the Church had not condemned the Copernican hypothesis, nor was it about to, because there was no fear that it could ever be proved to be necessarily true (Finocchiaro 1989, 303).

As the book neared completion, there were problems getting it approved by the Roman licensor, so Galileo had it approved in Florence. But when it was seen in Rome, it was greeted with dismay. Galileo's sarcastic preface placed the pope's reasonable concerns in the worst possible light; the treatment of the two systems was anything but balanced; and few doubted that it was intended to support the reality of the Copernican position in spite of formal disclaimers. Furthermore, throughout the text Galileo had placed the pope's arguments in the mouth of a character named Simplicio (Italian for simpleton). When the pope appointed a special commission to investigate the matter, they discovered the inquisitorial injunction that forbade Galileo from discussing the Copernican system at all. Urban VIII felt betrayed by his old friend on several levels. First, it seemed to him that Galileo had used his friendship to get around the Inquisition's 1616 injunction. Then, Galileo had written his dialogue in Italian rather than Latin to have maximum impact on an unsophisticated audience in spite of the concerns of the Church's leaders. Finally, he had made the pope into a laughingstock by identifying him with a simpleton.

Urban VIII seems to have become so angered that he entered into a personal vendetta to punish Galileo. He charged Galileo with violating the 1616 injunction. When an inquisitorial panel was presented by Galileo with the letter signed by Bellarmine, which did not prohibit discussion of the Copernican hypothesis, several members, including the pope's nephew, argued that under its own rules of procedure the Inquisition had to find Galileo's evidence the stronger. They recommended to the pope that Galileo simply be found guilty of using poor judgment. The pope would not agree and demanded that Galileo be examined to determine whether his intent had actually been to support the truth of the Copernican theory. On June 22, 1633, Galileo, having been found guilty of arguing for the truth of the Copernican position, and therefore of violating the weaker form of the earlier prohibition, was sentenced to recant his position and to be placed under arrest at the pleasure of the Inquisition.

After some negotiation, Galileo was allowed to remain for the remainder of his life under house arrest at his villa in Arcetri. There, in 1638, he produced his most important work in natural philosophy, *Discourses and Mathematical Demonstrations Concerning Two New Sciences Pertaining to Mechanics and to Local Motion*, which articulates the law that describes motion under the force of gravity. He died in 1642, within a few days of Isaac Newton's birth.

Once again we can ask whether what happened to Galileo is best understood as a fundamental conflict between science and religion. Or is it more appropriately understood as a personal conflict fueled by insensitivity and arrogance on the part of Galileo, by misunderstanding and misinformation regarding the events of 1616, and by injured pride on the part of Urban VIII? The pope did use the power of his office to see that Galileo was charged and punished; but misuse of power is neither confined to religious figures nor does it seem to most to be an essential feature of religion.

THREE ADDITIONAL SPECIAL CASES OF CONFLICT

Many of the episodes once labeled as conflicts between science and religion can more appropriately be seen as other kinds of conflicts; but there are three particular cases other than the Galileo case that deserve mention here because of their special significance for the tradition of writing about science and religion interactions. All three involved nineteenth-century developments, but all can easily be generalized to apply to other places and times.

Prior to the mid-twentieth century, the most often cited reason for the decline of nineteenth-century religious orthodoxy was that "because Lyell and Darwin had shown that neither the origin of the earth nor the origin of man as described in Genesis can be reconciled with the findings of science, therefore people became atheists and agnostics" (Murphy 1955, 800). On the contrary, argued Howard Murphy:

> Not only is this implausible on the face of it. It has also obscured the fact that the Victorian religious crisis was produced by a fundamental conflict between certain cherished orthodox dogmas (of which the infallibility of the letter was perhaps the least important) and the meliorist ethical bias of the age. Contemporary developments in geology, biology, and Biblical scholarship provided indispensable ammunition once the attack on orthodoxy was under way, but they did not generate the attack. (1955, 800–801)

Looking at the biographies of a number of Victorian intellectuals, it seems that most began as orthodox Christians but became disillusioned with their churches and with Christian doctrines when the churches and doctrines failed to conform to what they considered as Christ's central message—that each of us treat our fellow humans as we would be treated, with justice and compassion. A doctrine such as special election could not be reconciled with the mandate for justice and compassion, for example. And how could one support an Anglican Church whose clergy seemed to be opposed to any defensible sense of social justice?

Many of those who turned away from Christianity for such reasons subsequently turned to science to help them build a new secular substitute for traditional religion—a substitute that focused on helping humans in this world, rather than on salvation in an afterlife. This was the pattern followed by Richard Congreve, who abandoned his clerical appointment in the Anglican Church, turned to Comte's positivism, became the leader of the first positivist church in London, and became the nation's most outspoken critic of exploitative imperialist policies in India. It was also the pattern followed by his most famous congregant, Mary Ann Evans, a.k.a. George Eliot (1819–1880), whose novels promoted positivism. It was probably the path taken by T. H. Huxley (Desmond 1997). It was certainly the pattern followed earlier in Germany by two young theology students, David Strauss (1808–1874) and Ludwig Feuerbach (1804–1872), whose ideas will be discussed in chapter 6.

It would certainly be incorrect to say that involvement in science never led to an undermining of a person's religious faith. On the other hand, there seems to be very good evidence that at least in the eighteenth and nineteenth centuries it was probably more common for persons to have a crisis of faith grounded in apparent contradictions and conflicts within Christianity. Some then turned to the sciences to support their own ethical positions.

A second form of apparent Victorian conflict between science and religion began within the scientific community. Competition for control of such scientific institutions as the British Association for the Advancement of Science between young professionalizing scientists and an old guard of amateurs led by clergyman-scientists encouraged members of the young guard to insist upon a rigid separation of science from topics such as natural theology and from religious opinion in general. Their desire to garner more cultural authority for the institution of science also led the younger professionals into conflict with religious authorities in general (Turner 1978).

In this case, both sides claimed that there was a fundamental conflict between science and religion to strengthen their positions and generate public support. The real conflict, however, seemed to be between secular scientists and Christian clergy as they jockeyed for power, prestige, and control over social institutions and educational policies. Similarly, much of the fuel that fires the debate between professional biologists and advocates of creation science at the beginning of the twenty-first century comes from the desire of members of each group to impose either secular or religious control over public education in the United States. There are undoubtedly honest arguments on both sides linked to serious differences regarding the appropriate demarcation of science and religion; but what has made these arguments the center of a major political confrontation is their potential for mobilizing both religious fundamentalists and radical secularists in a battle for cultural authority.

A third source to support stories of conflict between science and religion emerged out of sectarian conflicts within Christendom, the usual claim being that some particular religious group should be viewed as inferior because of its opposition to unfettered scientific activity. Almost immediately after Galileo's trial, for example, Protestant authors began to argue that the outcome of the trial symbolized a Catholic authoritarian imposition of ignorance, as well as suppression of individual conscience, which was antithetical to Protestantism. This view was accepted by many mid-nineteenth-century Protestant intellectuals who were trying to assure themselves and their audiences that Protestantism was a friend to science. It was the view of John William Draper (1811–1882), an American chemist from a Protestant background, who published his *History of the Conflict between Religion and Science* in 1874. Though Draper's work had a lesser impact than White's, it was more popular in late nineteenth-century America, in part because of its overtly anti-Catholic orientation.

Quantitative studies done between the 1880s and the 1950s consistently showed that Protestants were significantly overrepresented among scientists relative to their numbers in European society. But it is not at all clear that differences in doctrine or dogma accounted for much, if any, of the differential participation rates. Part of the difference might have been a consequence of the greater educational emphasis on religion in Catholic regions. Another factor might have been the consequences of the practice of celibacy among Catholic priests, coupled with considerations of heredity. Scholars have suggested that many Protestant scientists were sons of clergy and may be presumed to have inherited outstanding intellectual abilities from their parent. On the other hand, Catholic priests, who probably represented the intellectual elite among Catholics, at least until very recently, did not (generally) produce offspring; so fewer Catholics had either the inclination or the inherited ability to become scientists (Cohen 1990, 145–150).

A similar intra-Christian battle that has had important consequences for the historiography of science and religion in America emerged in the mid-nineteenth century when a branch of New England Unitarianism became closely linked to the highly intellectual and highly speculative transcendentalist philosophy promoted by Ralph Waldo Emerson (1803–1882). Impatient with the "Baconian" methods associated with most American Protestant groups in the mid-nineteenth century, these advocates of a more "liberal" religion argued that evangelical and Calvinist religion "produced nothing but incoherent emotion and illiberal dogma" (Bozeman 1977, 165). Thus, they initiated a tradition of complaints that American Protestantism was anti-intellectual in general and anti-scientific in particular, which continued to inform writing on American intellectual history through the mid-twentieth century.

In the 1970s several historians of science initiated a reassessment of the relationship between American Protestantism and the sciences. While admitting that some strongly evangelical groups might reasonably be charged with anti-intellectualism and a negative attitude toward prioritizing scientific activities, these authors demonstrated that mainstream evangelical Protestantism in America promoted education, particularly scientific activities. Indeed, many American scientists and theologians alike in the early- to mid-nineteenth century viewed scientific study as a religious calling. The mainstream Protestant view of science was heavily dependent on the Scottish Common Sense philosophy, which emphasized "Baconian" induction as the proper source of scientific knowledge. In this emphasis, it stood in opposition to the highly speculative and intuitive strategies promoted by German idealism and American transcendentalism, and it represented an approach to the nature of both science and religion that was in diminishing favor among liberal theologians and scientists in the second half of the century. So the much touted conflict between evangelical religion and science in nineteenth-century America seems to have been more of a conflict between advocates of two radically different philosophies of knowledge promoted within two different religious traditions.

MODERN CLAIMS THAT RELIGION SUPPORTS SCIENCE

Discussions of the relation between Protestantism and religion were changed dramatically in the late 1930s when a sociology student, Robert Merton, argued that Puritanism, a particular branch of Calvinist Protestantism in England, was responsible for the leadership role England played in science in the seventeenth century. Puritanism promoted a complex group of sentiments that were just those needed to promote and guide the growth of modern science. Merton made it clear that it was not Puritan theology but rather the Puritan ethos—the core set of values that led men into certain paths of activity—that "established a broad base for scientific inquiry, dignifying, exalting,

[and] consecrating such inquiry" (Cohen 1990, 112). These values or sentiments included: an admiration for rational, tireless industry; a strong desire to glorify God "by a clear-sighted, meticulous study of his works"; an emphasis on social welfare, or "the comfort of mankind"; an exaltation of the faculty of reason—but a reason subordinated to empiricism; and a rejection of human authority in favor of the power of each individual to achieve his or her own understanding of both Scripture and nature (Cohen 1990, 113–126). They functioned primarily by directing persons "who hitherto might have turned to theology or rhetoric or philology into scientific channels," and by promoting a kind of science that was both utilitarian and experimental (Cohen 1990, 122).

This thesis has been rightly criticized on a number of grounds, for both its identification of the complex of values favorable to science as uniquely Puritan, or even as uniquely Protestant, and its identification of particular individuals, including Robert Boyle and John Wilkins, as Puritans. Nonetheless, Merton's work was doubly important. It initiated interest in the social dimensions of scientific practices, and it spawned an evermore sophisticated series of investigations of the positive interactions between religious and scientific activities, which will be summarized in subsequent chapters.

An important alternative to Merton's thesis was developed in a series of works written in the 1960s. In *Giordano Bruno and the Hermetic Tradition* (1964), Dame Frances Yates argued that the scientific revolution was deeply indebted to a Christian humanist tradition related to the recovery of a corpus of writings associated with Hermes Trismegistus, a supposed Egyptian wise man who pre-dated Moses by nearly a thousand years. In reality, the *Corpus Hermeticum,* translated into Latin by Marsilio Ficino in 1463, was a collection of third- or fourth-century neo-Platonic texts. But this was not known until the early seventeenth century. In the meantime it stimulated a Renaissance focus on arts such as alchemy, astrology, and natural magic. These arts were oriented toward active human participation in the governance of the world and constituted the leading edge of Renaissance science. In addition, according to Yates, Hermeticism, which emphasized the centrality of the sun, encouraged the acceptance of Copernican astronomy. Like the Merton thesis, the Yates thesis has generated both a justified critical response and serious investigations of the important place of the occult sciences in the scientific revolution.

What are we to make of all of these competing and often conflicting interpretations of support for science by religious groups stimulated by Merton's work and by that of Yates and her commentators? John Brooke, who is probably the most respected scholar working on science and religion interactions today, argues that we simply have to admit the complexity of the situation and acknowledge that interactions between religious and scientific groups can be supportive, conflicted, or neutral, depending entirely on the specific context in which the groups interact (Brooke 1991).

Chapter 2

Religion and the Transition to "Modern" Science: Christian Demands for Useful Knowledge

The character of dominant scientific knowledge and dominant scientific practices during the medieval period were radically different from those associated with modern science. With rare exceptions, medieval science was not intended to be applied—except to the understanding and appreciation of God's creation and as a background to medical training. Nor, in spite of a few exceptions, did it involve experimentation or even the direct observation of natural phenomena. It was all but exclusively theoretical and focused on book learning and verbal disputes. By contrast, modern science is characterized, especially in the public mind, with utilitarian goals. It is supported by society because it is expected to be applicable to national defense and to the improvement of the quality of life, both in terms of material wealth and in terms of improved health. Moreover, experimental work in laboratories and observational work in the field are central activities associated with producing scientific knowledge and with scientific education.

This chapter will begin by briefly characterizing late medieval science. Then it will concentrate on exploring the religious dimensions of the movement that produced a changed focus in the search for natural knowledge, from one that emphasized the contemplation of God's creation to one that emphasized the manipulation of the natural world in order to better the human condition. In this connection, it will consider the roles of Christian humanism and the recovery of the so-called *Hermetic Corpus* in promoting natural magic, which in turn stimulated the growth of the experimental sciences. It will consider the life and work of the medical alchemist Paracelsus as an illustration of how religion, the occult arts, and science interacted to mutually promote one another in the early modern period. Finally, it will consider the role of religious ideas in promoting a new institutional context for the pursuit of scientific knowledge through the writings of Tommaso Campanella, Johanne Andreae, and Francis Bacon and through the organizational work of men such as Samuel Hartlib.

THE STARTING POINT: LATE MEDIEVAL SCIENCE

Late medieval science was dominated by the writings of the Ancient Greek philosopher Aristotle; so Aristotelian ideas need to be briefly characterized in order to set the stage for the transformation of scientific knowledge and practices that occurred in early modern Europe.

Aristotle, who lived much of his life in Athens where he died circa 323 B.C.E., was, with his teacher Plato, one of the two greatest philosophers of antiquity. He was also one of the greatest scientists of all time—a view shared by William Harvey and Charles Darwin. The promoter of an empirically grounded approach to knowledge, Aristotle wrote treatises on a wide range of topics that defined curricular areas that remain central to academic life even today. His *Physics* defined the domain of natural philosophy, what we called physics again in the twentieth century. *De Caelo et Mundo* (*On the Heavens and the Earth*) defined the domain of astronomy; *De Anima* (*On the Soul*) covered what we would now call psychology. *Prior Analytics* and *Posterior Analytics* defined the domains of logic and the theory of knowledge. *Nichomachian Ethics* and *Metaphysics* defined continuing domains within philosophy, and his works on living beings established the foundations for comparative anatomy, zoology, and ecology.

Aristotle's way of dividing knowledge into three categories played a significant role among medieval Christian scholars. His *productive knowledge* was manual work like carpentry. As a teacher of the Greek landowning class, Aristotle disdained productive knowledge as the lowest form of knowledge. It was beneath his students' concern, he argued, because it was, for the most part, learned by rote, without an understanding of why it worked. On the other hand, both *theoretical knowledge*—that is, pure contemplation of things that one cannot change—and *practical knowledge*—which guides ethical and political action—offer the opportunity for humans to exercise their reason, which is what makes them superior to other beings. Of these two, contemplation is the higher activity for several reasons, the most important of which is that humans have within them a small element of the divine. This divine portion is very small but "it far surpasses every thing else in power and value" (Aristotle 1177b27–1178a4). And, according to Aristotle, "the activity of the divinity which surpasses all others in bliss must be the contemplative activity" (Aristotle 1178b20–24). He continues:

> As a consequence, a human's greatest happiness must lie in contemplation: The Gods enjoy a life blessed in its entirety; men enjoy it to the extent that they attain something resembling the divine activity; but none of the other beings can be happy, because they have no share at all in contemplation, or study. So happiness is coextensive with study, and the greater the opportunity for studying, the greater the happiness, not as an incidental effect, but as inherent in contemplation. (Aristotle 1178b25–31)

Those who sought to replace Aristotelian approaches to science in the early modern period were equally—or even more—insistent on the divine element in man (though not necessarily in woman). But they viewed God's creation of the universe as vastly more admirable than mere contemplation. For these men, the highest exercise of their human faculties lay not simply in understanding God's creation, but in acting in the world either to complete God's unfinished work or to transform it to serve human ends.

Aristotelian theoretical knowledge had two important characteristics associated with the fact that it focused on things that humans cannot change. The best practical knowledge could never be certain. Politics and ethics were thus about what usually happens rather than about what always happens, and their claims were always to some degree uncertain. Theoretical sciences, on the other hand, were about what always or almost always happens—we say they were deterministic—and their knowledge claims were capable of the highest degree of certainty. In fact, if the basic assumptions or "postulates" of an Aristotelian theoretical science were unquestionable, the results derived from them were also unquestionable. The ancient Greeks formulated Euclidean geometry to meet the criteria of Aristotelian theoretical science, and it stood as an example of the level of certainty that many medieval scholars hoped to achieve in the natural sciences.

Finally, Aristotelian theoretical knowledge claims always involved four kinds of "causes" of phenomena: material, efficient, formal, and final. Consider, for example, a house. Its material cause is the bricks, mortar, boards, and so forth, which enter into its construction. Its efficient cause is the masons, carpenters, and so on, who do the work. Its formal cause is the idea that the contractor has about its structure or the blueprint that establishes its form. Its final cause is the reason for which it is being built—that is, the desire of some person or persons to have a home. Any science that emphasizes the end for which something is done is said to be *teleological,* and Aristotle's natural or theoretical sciences were all deeply concerned with such ends or final causes.

When Aristotelian knowledge re-entered Europe in the Middle Ages, it seemed incompatible with the then prevailing character of Christianity, which owed much to the thought of Plato. But by the end of the thirteenth century, Aristotelian science and Catholic Christianity had, for the most part, accommodated each other. Some major features of the resulting Christianized Aristotelian view of the world are described below so that challenges to them can be understood later.

According to Aristotle, the universe as a whole was not self-moving. It was spherically shaped and rotated once a day, moved by what Aristotle had described as the prime unmoved mover and what Christian scholars almost universally identified as God, who thus stood outside of the physical universe. Inside the sphere of the prime mover, or *primum mobile,* was the sphere of the fixed stars, and nested inside that were the spheres of the various planets as they were then identified—Jupiter, Saturn, Mars, the Sun, Venus, Mercury—and the Moon.

Inside of the lunar sphere, at the greatest distance from that of the prime mover, were the spheres of the four terrestrial elements: fire, air, water, and (at the very bottom, or center) earth. The natural motion of each of these elements was toward or away from the center. Moreover, each of the terrestrial elements was said to be "corruptible"; that is, each could be "transformed" into one of the others either by wetting, drying, heating, or cooling. Earth was dry and cool, water was wet and cool, air was hot and wet, and fire was hot and dry, and each could be transformed by changing one of its features.

This Aristotelian view of the world corresponded very well to the common experiences of ordinary people. While the heavenly bodies all seemed to move around the earth with a daily rotation, each of the wandering stars or planets seemed to have its own motion through the fixed stars. The earth certainly feels solidly fixed beneath our feet. While it takes effort to force ordinary bodies on the surface of the earth to move horizontally, heavy bodies such as rocks naturally fall downward in the atmosphere, and light bodies such as air naturally seem to rise when they are in water. When wood burns, it seems that part is converted into flames (fire), which appear to strive to move upward through the air. Similarly, air seems to be transformed into water in clouds, and when it is, it falls downward toward the earth as rain.

Aristotle had not discussed interaction between the portion of the universe above the sphere of the moon and that below in any detail, but he did suggest that somehow, just as the sun governed the growth of plants, other planets and stars might govern other terrestrial phenomena (Aristotle *On Generation and Corruption*, 336b). Islamic and medieval European astrologers and medical practitioners grafted a set of ideas on to this Aristotelian notion that was very widely accepted both among scholars and among laypersons. It was generally agreed that sublunar events and entities could have no effect on what went on above the lunar sphere. But it was also agreed that both constellations of stars and individual planets in some sense regulate or "govern" terrestrial events by giving off substances—literally "influences," because they flowed inwards—that could have an effect on earthly phenomena.

An extremely important doctrine in late medieval Aristotelian science that came into Europe from Islam was that of correspondences or of macrocosm-microcosm parallels (see Figure 2.1). According to this doctrine, as illustrated in the table of Figure 2.1, God created the universe in such a way that every level of existence was structured to mirror every other level of existence, and all levels were hierarchically ordered. For example, just as the sun was the most noble of heavenly bodies, gold was the most noble member of the order of metals, the laurel was the most noble of trees, and the lion was the most noble member of the animal kingdom. At the other end of the value spectrum, just as lead was the least noble metal, the ass was the least noble of animals, and hypochodria the least noble of illnesses.

The doctrine of correspondences was usually combined with the theory of astrological influences to identify the influences from certain star constellations and planets with certain portions of the human body—the moon with the head, for example, which is where our no-

Tabula rerum naturalium, quæ ſingulis ſubijciuntur Planetis.

	☼	☽	♂	☿	♃	♀	♄
Metalla.	*Aurum.*	*Argentum.*	*Ferrum.*	*Cuprum.*	*Aes.*	*Stamnum.*	*Plumbum.*
Lapides pretioſi.	Carbunculus.	Corallus, Calcedonis.	Magnes, Hyacinthus, Amethiſtus.	Achates, Chryſolithus.	Saphirus, Smaragdus.	Turcheſia., Margarita, Berillus.	Onix, Iaſpis, Topazius.
Arbores.	Laurus, Cinnamomum.	Pomus.	Quercus.	Mali punici.	Citrus, Ficus.	Myrthus, Vitis, Olea.	Taxus, Cypreſſus, Meſpilus.
Grana.	Lupinus, Caltha.	Colutea.	Piper.	Grana Chermes.	Grana Ben.	Grana Pinorũ.	Gith.
Herbæ, flores.	Heliotropium, Hypericon.	Lunaria, Ranunculus, Artemiſia.	Verbena, Iſatis, Glaſtũ, Ruta, Abſynthiũ.	Hyacinthus, Narciſſus.	Ænula, Valeriana, Betonica.	Satyria, & Orchides.	Helleborum, Aconitum, Napellus.
Animalia.	Leo, Gallus.	Ælurus, Grus.	Lupus, Taurus, Accipiter.	Canis, Hirundo.	Equus, Aquila.	Ceruus, Columba, Bufo.	Aſinus, Noctua.
Colores.	Splendidus, lucidus.	Candidus.	Flammeus.	Ferrugineus, Cinericius.	Rubeus.	Viridis, cœruleus.	Fuſcus, plumbeus.
Morbi.	Morbi calidi.	Morbi à frigiditate, & humiditate orti.	Febris ardens, Gangręna, Cãcer.	Apoplexia., Philomania.	Defectus ſpirituum, Cardialgia.	Gonorrhæa, Satyriaſis.	Hypochõdriacus affectus.

Atque hæc ſunt, quæ de hoc Sciatherico phyſico-medico-mathematico dicenda duximus. Synopſim verò rerum, quas continet, ſequenti tetraſtico comprehendimus.

Yyy 2 *Abdita*

Figure 2.1. Table of Correspondences from Athanasius Kircher's *Mundus Subterraneus* (1665). To each planet (including the sun and moon) corresponds a particular metal, precious stone, tree, grain, herb, animal, color, and disease in this table. Other similar tables might also include body parts and social roles. Courtesy of Special Collections, Honnold/Mudd Library.

tion of "lunacy" comes from. This whole system was often depicted visually in late medieval and early modern scholarly works and in almanacs for laity as well (see Figure 2.2).

CHALLENGES TO MEDIEVAL SCIENCE

Two important challenges emerged to mainstream Aristotelianism within medieval culture. Each was associated with a theological position that was declared heretical, and neither became dominant until the Renaissance. The first was the rebirth of an early Christian emphasis on the second coming of Christ, attended with a concern with

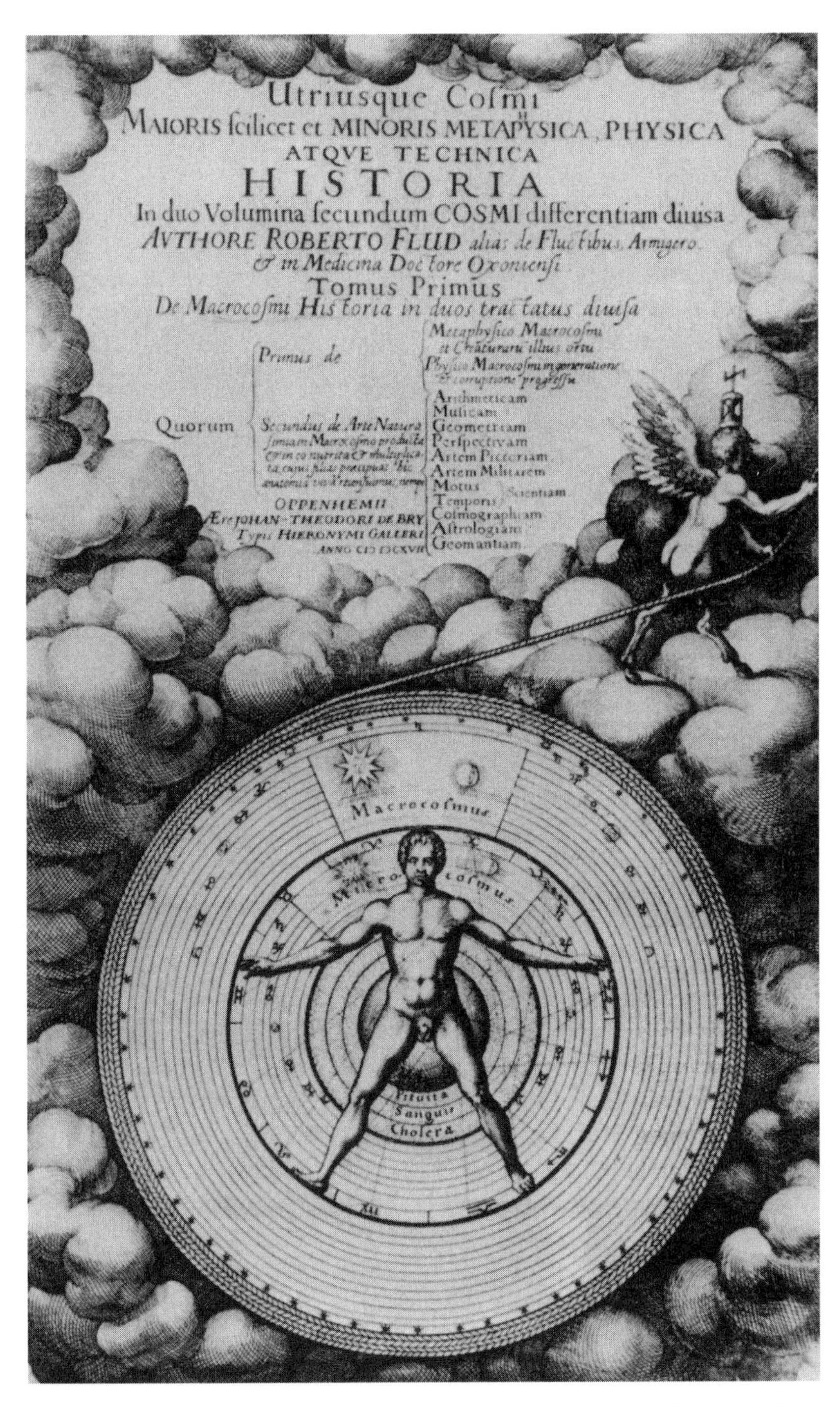

Figure 2.2. Title page from Robert Fludd's *Ultrisque Cosmi . . .* (1618). Note the lines that link specific signs of the Zodiac to specific parts of the body.

what the prophetic books of Daniel and Revelations said needed to be done to prepare for this return. People who looked forward to the returned Christ's thousand-year reign on earth were called "millenarians." They turned against the contemplative focus of Aristotle's science because the prophetic books of the Bible called for a knowledge that could help perfect the world.

Beginning as early as the eighth century, Christian monasticism began to abandon the aristocratic emphasis on theoretical knowledge that had come down from classical antiquity and which Aristotelian sources would reinforce. Monastic communities had to be self-sufficient, which meant that monks had to engage in productive labor. And a theoretical justification of the dignity of human productive labor and of the productive arts soon emerged. The "mechanic arts," such as agriculture, hunting, and food preparation, were linked to the notion that God created man in his image and likeness and that this likeness was manifested in the human capacity to rule the things of the earth. With the Fall, man lost his special role as ruler of the earth. But the mechanic arts have been given to man to restore his ability to rule (Ovitt 1986, 120).

For Joachim of Fiore (c. 1135–1202) and his followers, the restoration of man to his condition before the Fall was the central focus of religious concern, for it signaled the thousand-year reign of Christ on earth foretold in the Book of Revelation. That thousand-year reign would restore the earthly paradise into which Adam was placed, and it was the necessary prelude to the ascension of the souls of the righteous into Heaven. Furthermore, for some followers of Joachim, including the fourteenth-century Franciscan scholar Roger Bacon (1214–1292), natural knowledge was understood to be the direct cause of the perfection of the world and the means by which the forces of good would overcome the forces of the Antichrist.

The expectation that Christ's second coming was imminent and that it would create the perfection of the world had been declared heretical in 431 on the grounds that the establishment of the Church had already initiated Christ's reign on earth, and it remained a heresy throughout the Middle Ages. So Joachim's and Roger Bacon's millenarian views were formally discouraged by church officials. Joachimite ideas did spread into popular culture and among radical Franciscans; but it was not until the humanist movement of the fifteenth century that they spread rapidly within the intellectual elite of Europe.

The second major challenge to medieval science was linked to an extreme and one-sided focus on God's omnipotence. This emphasis had intensified among theologians who saw in Aristotelian claims about the necessary character of theoretical science a challenge to God's absolute freedom to create any kind of world that he might please.

During the thirteenth century, as Aristotelian texts came increasingly to dominate the liberal arts in the medieval university, a number of conservative theologians began to publish lists of "errors" made by Aristotle and his commentators. In general, any of Aristotle's claims that seemed to conflict with a widely held Christian principle or to be inconsistent with some biblical passage was declared an error. Far and away the most common reason for declaring an Aristotelian claim to be an error, however, was that it violated the principle of God's absolute omnipotence by limiting his ability to act in some arbitrary fashion. Consider the following example from Giles of Rome's *On the Errors of the Philosophers, Aristotle, Averroes, Avicenna, Al Gazel, Al Kindi, and Maimonedes*, written around 1224:

> Because he [Aristotle] wished to proceed by way of nature, he believed, as is clear from book 4 of the *Physics*, that since two bodies cannot *naturally* be in the same place, it was so essential for dimensions to resist dimensions, that it is impossible for dimensions to continue to exist and yet not resist dimensions.
>
> Because of this it follows that God could not make two bodies exist in the same place.

According to Scripture, however, God is omnipotent, and thus able to make two bodies exist in the same place if he wishes to do so, proving the error of the Aristotelian claim.

In the writings of William of Ockham, the so-called "voluntarist" theology (because it emphasized God's free will), which underlay such an identification of errors, developed into a philosophy called nominalism. One of the central features of nominalism was the claim that since God's will is the final cause of all things, and since his will is ultimately unfathomable by humans, we must give up the teleological emphasis of the Aristotelian sciences. Indeed, because God could choose to cause any event in any way, we must give up seeking "causal" knowledge in the Aristotelian sense in favor of describing how events occur.

Since God is omnipotent and his intentions unknowable, all scientific knowledge that makes claims about existing phenomena must be contingent rather than necessary. That is, knowledge of the natural world depends upon the experiences we have of that world and can not be derived rationally prior to that experience. We say today that scientific knowledge must have an empirical foundation. Furthermore, experiences cannot produce knowledge claims that are necessarily true.

Finally, nominalism did allow that certainty was attainable in theoretical systems that began from some set of postulates created by an author and that were assumed to be true for purposes of an argument. If one derived consequences from those postulates using nothing but logic, those consequences must be true within the context of the argument. For example, when the French theorist Nicole Oresme posited the existence of bodies that undergo "uniformly diform motion"—that is, motion that increases its speed in a straight line by equal amounts in equal times—he was able to prove that in any given time they would pass through a distance proportional to the square of the time. What he could not do was prove that any such bodies actually exist.

Many of the characteristics of modern science emerged out of nominalist philosophy. These include the abandonment of attempts to answer the question of why things happen in favor of describing how they happen, usually through mathematical "laws." Modern science also shares with medieval nominalism the belief that scientific knowledge claims can only be legitimate if they are derived from and tested against experience. Thus, in an ironic way, a conservative religious argument gave a strong impetus to modern ways of doing science. But like millenarianism, Ockham's philosophy was resisted by most medieval thinkers, not least because it undermined the very possibility of a science of theology. Only with the humanist movement of the fourteenth and fifteenth century did nominalism achieve a central position within notions of how we attain knowledge.

CHRISTIAN HUMANISM AND THE *HERMETIC CORPUS*

During the late fourteenth century, beginning in the universities of Northern Italy, an educational reform movement that challenged the primacy of logic and Aristotelian natural philosophy began to develop. Scholars associated with this movement promoted a much greater emphasis on the *studia humanitatis*, especially grammar, rhetoric, and poetics, and thus came to be called humanists. The emergence of humanism was closely connected with the vital civic culture of Northern Italy, a culture in which involvement in both the economic and political life of one's city was becoming much more highly valued than the contemplative life.

Aristotelian natural philosophy and logic were relatively useless as preparation for men of affairs, who were much more concerned with effective ways of presenting arguments. Early humanists such as Francesco Petrarch, argued that skill in rhetoric, which involved not merely techniques of delivery, but the wisdom and judgement which came from the study of the great Roman orators—especially Cicero—was much more important as a foundation for success as a citizen than theoretical knowledge of nature. They thus promoted a return to the elegance of classical Latin, as opposed to what they viewed as a corrupted medieval Latin. In addition, they fostered the recovery of classical texts on politics, morals, history, and so on, which had not been preserved within the European monastic culture largely because they represented pagan rather than Christian values. Lest anyone fear that humanism had pagan rather than Christian intentions, Petrarch assured them that this was not so:

> Christ is my God; Cicero is the prince of the language I use. I grant you that these ideas are widely separated, but I deny that they are in conflict with each other. Christ is the Word and the Virtue, and the Wisdom of God the father. Cicero has written much on the speech of men, on the virtues of men, and on the wisdom of men—statements that are true and therefore, surely acceptable to the God of Truth. (1910, 19)

Finally, early humanists looked upon the practical sciences as more certain than the natural, or theoretical, sciences precisely because they did depend on human rather than divine action. Borrowing from nominalism, they argued that certain knowledge about the natural world is forever beyond our reach because natural phenomena ultimately depend on God's unknowable will. On the other hand, legal codes and political arrangements can be understood because they are human creations.

Humanism spread rapidly beyond Italy, and humanistic interests expanded rapidly beyond a focus on Roman oratory. Soon humanist scholars were seeking examples of Greek eloquence and wisdom, such as that of the dialogues of Plato, which were known—except for the *Timaeus*—only by reputation in Europe. Theologians with humanistic backgrounds began to argue that the highly theoretical orientation of scholastic Aristotelian theology might be a corruption of an earlier and purer Christianity that could be discovered by recovering ancient religious texts. So Hebrew joined Greek and Latin as languages that humanists learned and used to translate documents previously unknown in Europe. Furthermore, scholars began to search for a *prisca theologia*—an original theology—which they began to think may have pre-dated even Moses. Wealthy patrons, such as the Medici family of Florence, sponsored expeditions to scour monasteries to locate ancient texts, and they funded humanist scholars to translate those texts into elegant Latin so that whatever wisdom they might contain could be recovered.

Around 1460, Marsilio Ficino, a humanist with strong theological and medical training, was working for the Medicis translating a series of Platonic dialogues that had been unearthed by family agents. Then in 1463, an exciting new set of documents arrived in Florence from a Macedonian monastery. These documents were to have a major impact on Ficino's career and on both religion and science. Known as the *Hermetic Corpus*, they were purportedly the writings of an ancient Egyptian wise man who had been admired by early church fathers including Clement of Alexandria. Moreover, they were believed to predate Moses by nearly a thousand years, and they contained doctrines that seemed to anticipate not only the core ideas included in Plato's *Timaeus*, but also the central doctrines of Moses and of Christianity.

Nearly 150 years later, Isaac Casaubon demonstrated that these texts were actually from the fourth century C.E. and were written by several authors from a number of different theological perspectives. But when they arrived in Florence they appeared to be by a single, extremely ancient author whose mystical experiences gave him access to the foundations of Judaism and Christianity—perhaps to the *prisca theologia* itself. In addition, it appeared as though both Moses and Plato had been drawing from hermetic sources; so it seemed absolutely critical to make their contents available. Ficino abandoned his work on Plato to translate the *Hermetic Corpus* and to weave the various short texts together into a more or less coherent body of thought.

Though they might seem to modern Christians to be quite heretical, the Medici Pope, Alexander VI, indirectly declared the core Hermetic ideas consistent with church doctrine in 1493 when he found Ficino's follower, Pico della Mirandola, innocent of wrongdoing in perpetuating many hermetic views. The number and distribution of copies of Ficino's translation indicate that the *Hermetic Corpus* was studied throughout both Catholic and Protestant Europe during the sixteenth century. Finally, from the frequency with which Hermetic doctrines were cited and alluded to, it is clear that they were widely admired among humanist intellectuals.

Several features of the Hermetic works are important. They begin with the central character, Hermes, having a vision in which a blinding light appears, followed by a thundering voice. He is told that the light is God and the voice that speaks the holy word is the son of God, who acts together with him to create the physical universe. This vision is consistent with the widely held early Christian notion that Christ was the *Logos*, or Word, and the co-creator of the world with God. The world is then created in a way that fuses elements from Plato's *Timaeus* and the biblical Genesis stories.

First, the Word creates the seven planets, who, as in the *Timaeus*, serve as governors of the sensory world. Then, "the Maker worked together with the Word, and encompassing the orbits of the Administrators, and whirling them round with a rushing movement, set circling the bodies he had made and let them revolve" (Hermes Trismagistus 1993, 119). Next, the birds, fish, four-footed, and crawling creatures are created by nature but in the sequence described in Genesis. Finally, as in Genesis, God created man in his own likeness, and because God delighted in man, he "delivered over to man all things that had been made" (Hermes Trismagistus 1993, 121). At this point, because the planets also love the likeness of the divine in man, each grants to man a share of its own nature—that is, some of its ability to govern the world and influence its phenomena—as man descends from the heavens to the earth. For these reasons, "man, unlike all other living creatures upon the earth, is twofold. . . . He is immortal [by virtue of his likeness to God], and has all things in his power; yet he suffers the lot of a mortal, being subject to Destiny" (Hermes Trismagistus 1993, 123).

This passage emphasizes the governing as opposed to the merely contemplative role of man. In that connection, it encourages a new, activist understanding of astrology, suggesting that man can not only use his astrological knowledge for predictive purposes, but also for purposes of intervening in the world to accomplish his own ends. Hermeticism, as a form of Christian humanism thus reinstituted a concern with natural knowledge into humanism. But in doing so, it sought a very different kind of natural knowledge than that of the scholastic followers of Aristotle. Over and over within the Hermetic writings there is an insistence that man is both mortal and divine and that he is intended both to admire the creation and to "administer" it. Without explicitly referring to millenarian doctrines, the Hermetic writings powerfully reinforced the notion of Joachim of Fiore and others that man was intended to do more than contemplate. Moreover, his activity in the universe was to be more than political. Man was to act in the universe by using his knowledge of nature and those powers granted to him by God and by the planets to improve the human condition—which would complete and perfect the creation.

Two additional features of the universe as portrayed in the *Hermetic Corpus* were of importance in the development of the sciences. First, unlike Aristotle's universe, in which the cause of motion lay outside the physical world, the Hermetic universe, like Plato's, was alive and self-moving, guided by what came to be called the "world-soul." The entire cosmos could thus be thought of as an immortal living being or a great animal. Many scientists who later rejected the notion of a living universe appealed to religious concerns. They argued that a living universe did not seem to need God's continued presence to function and that it therefore undermined belief in the continuing dependence of everything upon God. Furthermore, they argued that there was a danger that men would become confused and worship the creation rather than the creator.

Secondly, sections of the *Hermetic Corpus* single out the sun as a uniquely important portion of the universe. At one point Hermes is told, "all things are filled with light . . . which is shed on all below by the working of the Sun; and the Sun is the begetter of all good, the ruler of all ordered movement, and the governor of the seven worlds [planets]" (Hermes Trismagistus 1993, 213). At another, he is told, "The sun is the greatest of the Gods in heaven; as to him all the gods of heaven yield place" (1993, 159). Some scholars believe that this emphasis on the sun made those who were influenced by hermetic doctrines more open to heliocentric (sun-centered) astronomical theories (Yates 1964).

There is little doubt that the popularity of Hermetic doctrines promoted the study of a number of non-Aristotelian knowledge traditions, which formed foundations of the modern sciences. None of these traditions were uniquely or even primarily Hermetic, but they gained authority by virtue of their connection with the primary values expressed in the Hermetic writings. Chief among these traditions was one associated with the terms "natural magic" and "occult arts."

In 1489, Ficino published *De Vita Coelitus Comparanda*, perhaps the most influential Renaissance book on magic. In it he appeals to a wide range of ancient philosophers and magi, including Zoroaster, Pythagoras, Democritus, Plotinus, and Proclus, as well as Hermes, in defending his natural magic. Natural magic was distinguished from and opposed to demonic magic, which was purportedly used for evil purposes and was condemned by the church. Ficino describes the use of natural magic as follows:

> Using natural objects, natural magic captures the beneficial powers of the heavenly bodies to bring good health. This means of action must surely be conceded to those who use their talents lawfully, just as it is in medicine and farming. . . . One practices agriculture, another mundiculture. . . . Just as the farmer tempers his fields to the weather and to give sustenance to man, so this wise man, this priest, for the sake of man's safety tempers the lower objects of the cosmos to the higher. . . . [Natural magic] puts natural materials in a correct relationship with natural causes. (Copenhaver 1990, 280–281)

Such a natural magic is possible because the natural connections among terrestrial and celestial objects can be understood and manipulated by those who have chosen to exercise that divine capacity for knowledge and action granted to man by God and by the planets.

Most of the connections and powers used by the natural magician were said to be "occult," or hidden, because they could not be seen directly. Instead they had to be discovered using some kind of schema. The most common schemas were the theory of signatures, which said that God placed "signs" within natural objects to indicate what they were useful for (see Figure 2.3), the theory of macrocosm-microcosm correspondences, and the Jewish Cabala, which involved interpreting the numerical values of the letters in the word for an entity or phenomenon.

Sometimes, however, empirical techniques lay at the foundation of an occult art. For example, much of alchemy, which was among the occult arts encouraged by Christian humanists, was grounded in direct observation of nature and in the development of laboratory procedures (see Figure 2.4). Describing alchemy, the fourteenth-century practitioner, Bonus of Ferrara wrote:

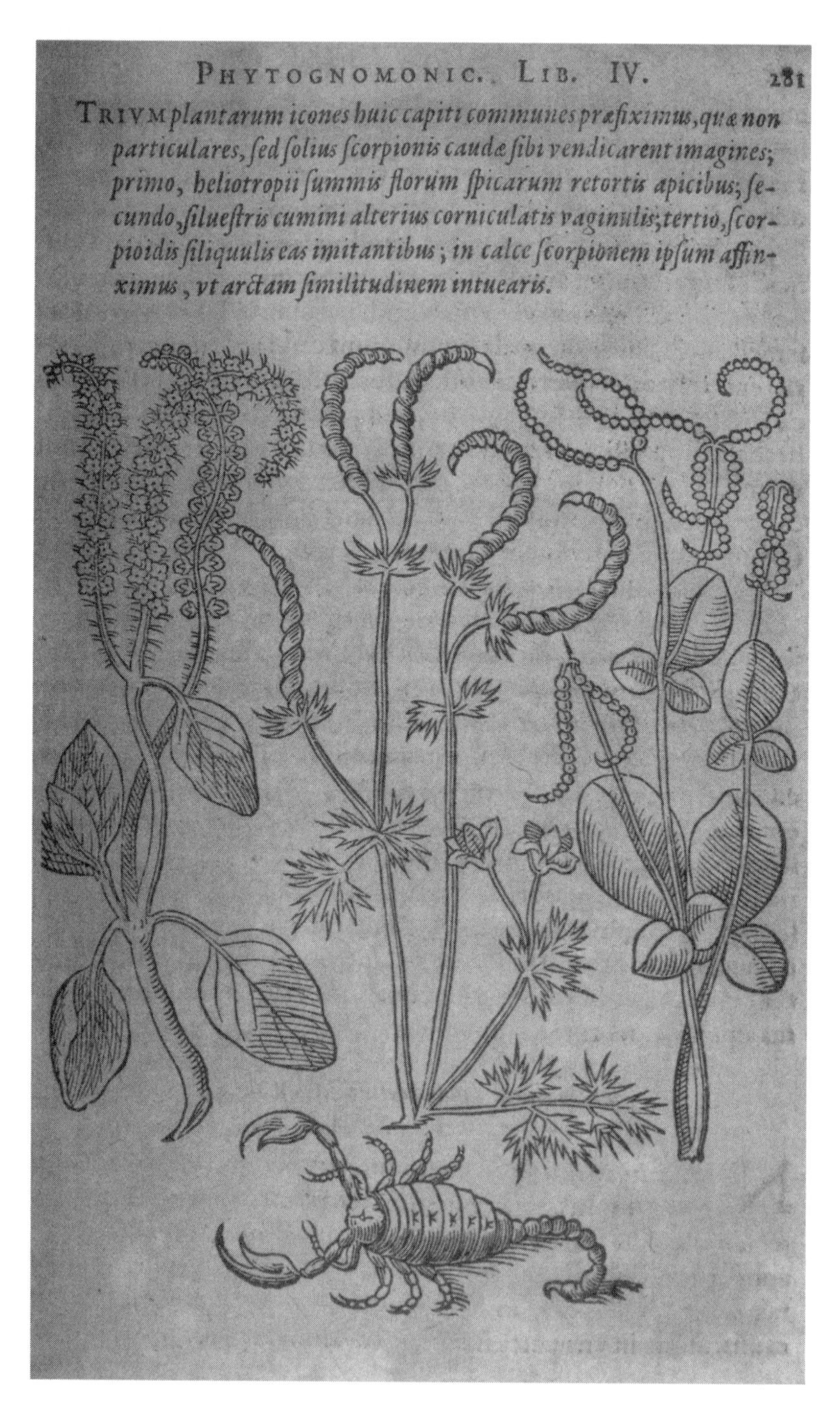
PHYTOGNOMONIC. LIB. IV. 281

TRIVM plantarum icones huic capiti communes præfiximus, quæ non particulares, ſed ſolius ſcorpionis caudæ ſibi vendicarent imagines; primo, heliotropii ſummis florum ſpicarum retortis apicibus; ſecundo, ſilueſtris cumini alterius corniculatis vaginulis; tertio, ſcorpioidis ſiliquulis eas imitantibus; in calce ſcorpionem ipſum affinximus, vt arctam ſimilitudinem intuearis.

Figure 2.3. Page from Athanasius Kircher's *Ars Magna Lucis Et Umbrae In Decem Libros Digesta*. The shape of the flowers or seed pods of the herbs are "signs" that they can be used to cure scorpion stings because they mimic the shape of the scorpion's tail. Courtesy of Special Collections, Honnold/Mudd Library.

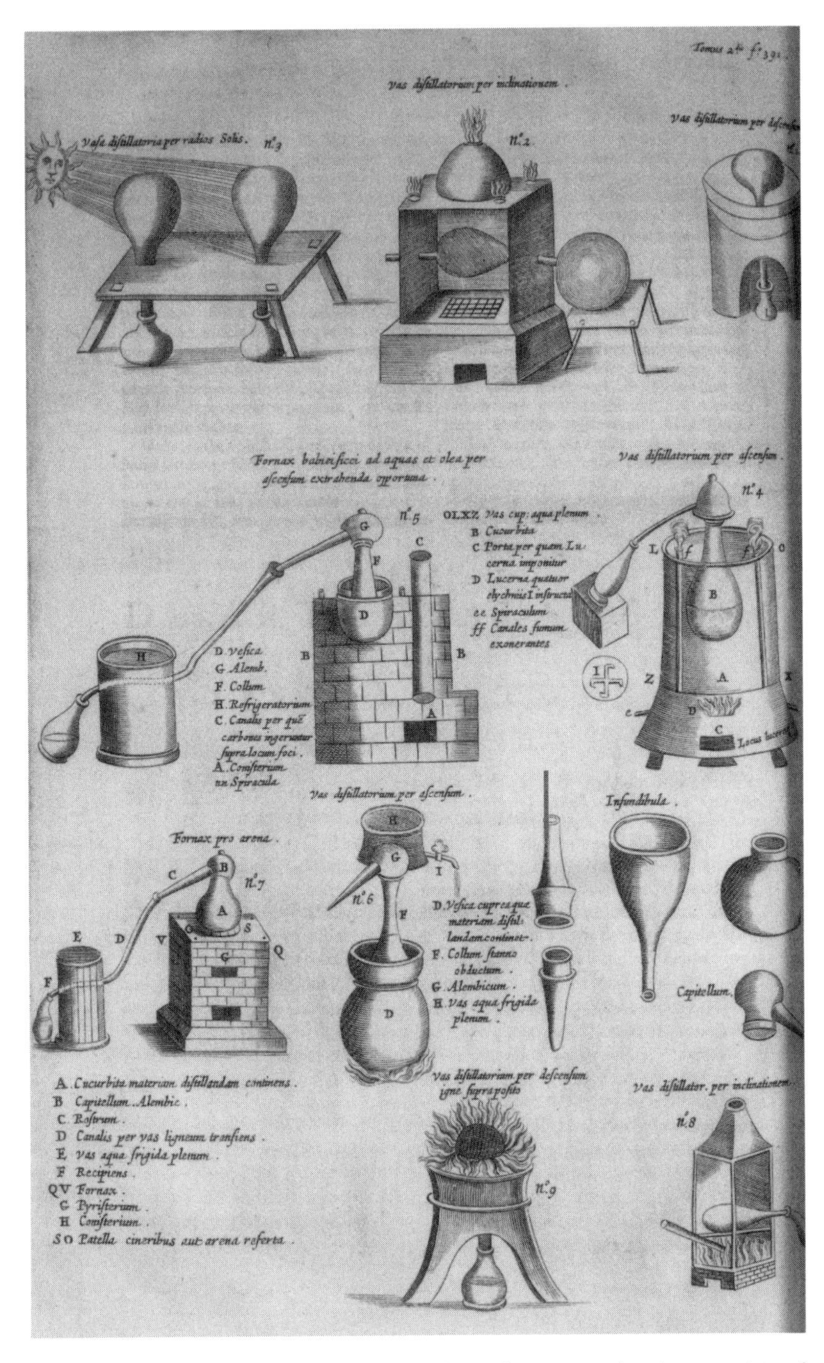

Figure 2.4. Diagrams from Kircher's *Mundus Subterraneus* showing a variety of devices for distillation, one of the most common alchemical processes. Courtesy of Special Collections, Honnold/Mudd Library.

> If you wish to know that pepper is hot and that vinegar is cooling, that colocynth and absinthe are bitter, that honey is sweet, and that aconite is poison, that the magnet attracts steel, that arsenic whitens brass and that tutia turns it of an orange color, you will, in every one of these cases, have to verify the assertion by experience.... A like rule applies with double force in alchemy, which undertakes to transmute the base metals into gold and silver.... The truth and justice of this claim, like all other propositions of a practical nature, has to be demonstrated by a practical experiment, and in no other way can it satisfactorily be shown. (Debus 1978, 17–18)

Another occult art fostered by some Christian humanists was what Tomasso Campanella called "real artificial magic" and what John Wilkins, one of the founders of the Royal Society of London, called "Mathematical magic," in a book by that title. Some mathematical magicians, such as John Dee, were primarily concerned with numerology, but others, including Campanella and Wilkins understood what we would now call simple mechanics as a form of mathematical magic.

It would be incorrect to suggest that the tradition of mathematical mechanics deriving from Archimedes and Hero of Alexandria owed its existence to Hermetic motives. It had already been resurrected in the thirteenth century and was growing increasingly powerful among completely secular scholars and engineers. But Christian humanist attitudes almost certainly expanded interest in mathematical works in the early modern period. It was, for example, the Hermetic magician, or magus, John Dee, perhaps best known for his conversations with angels, that produced the first English translation of Euclid's *Elements*.

Just as Christian humanism increased interest in subjects such as astrology, astronomy, alchemy, and—to a lesser extent—mathematics, it promoted academic scholarly interest in a wide range of practical subjects, from natural history, through various crafts such as metallurgy, dyeing, and so forth, to mining. At one end of the spectrum, Roman encyclopedist Pliny the Elder's *Natural History* was published numerous times during the sixteenth and seventeenth centuries (see Figure 2.5). This work described various plants, animals, and minerals, offered comments on their uses (including the magical use of mineral talismans), and proffered practical advice on such topics as gardening, viticulture, and animal breeding.

At the other end of the spectrum, Georg Bauer, who took on the Latin name Georg Agricola for literary purposes, published *De re*

THE

HISTORIE

OF THE WORLD:

Commonly called,

THE NATVRALL HISTORIE OF

C. PLINIVS SECVNDVS.

Tranſlated into Engliſh by PHILEMON HOLLAND
Doctor of Phyſicke.

The firſt Tome:

LONDON,
Printed by Adam Iſlip.
1634.

Figure 2.5. Title page from the 1634 English edition of Pliny's *Natural History*. One of the first moves of many humanists was to replace Aristotle with Pliny as the principal source of natural philosophy because of Pliny's focus on applied knowledge. Courtesy of Special Collections, Honnold/Mudd Library.

Figure 2.6. Illustration of the appearance of veins of ore in a hillside from Georg Agricola, *De re Metallica* (1912 English translation by Herbert Hoover of 1556 Latin original). This book was left on the altars of New World churches so that miners would come into the church when they needed to consult it. Courtesy of Special Collections, Honnold/Mudd Library.

Metallica in 1556. This textbook on mining technology in elegant Latin is famous for its magnificent drawings of geological formations and mining devices (see Figures 2.6 and 2.7). It stands as an example of the new scholarly attempt to make public what had been the proprietary knowledge of craft guilds and other specialist groups previously disdained by most scholars.

Especially in the Germanies, where the University of Wittenberg provided a model for other institutions, the Christian humanist emphasis on applied natural knowledge was embodied in reforms of the scientific portion of the curriculum. Phillip Melancthon (Greek for his birth name, Schwartzerd), who was Martin Luther's closest colleague and a learned humanist scholar, insisted in the 1520s that natural philosophy be taught at Wittenberg from Pliny's *Natural History*, rather

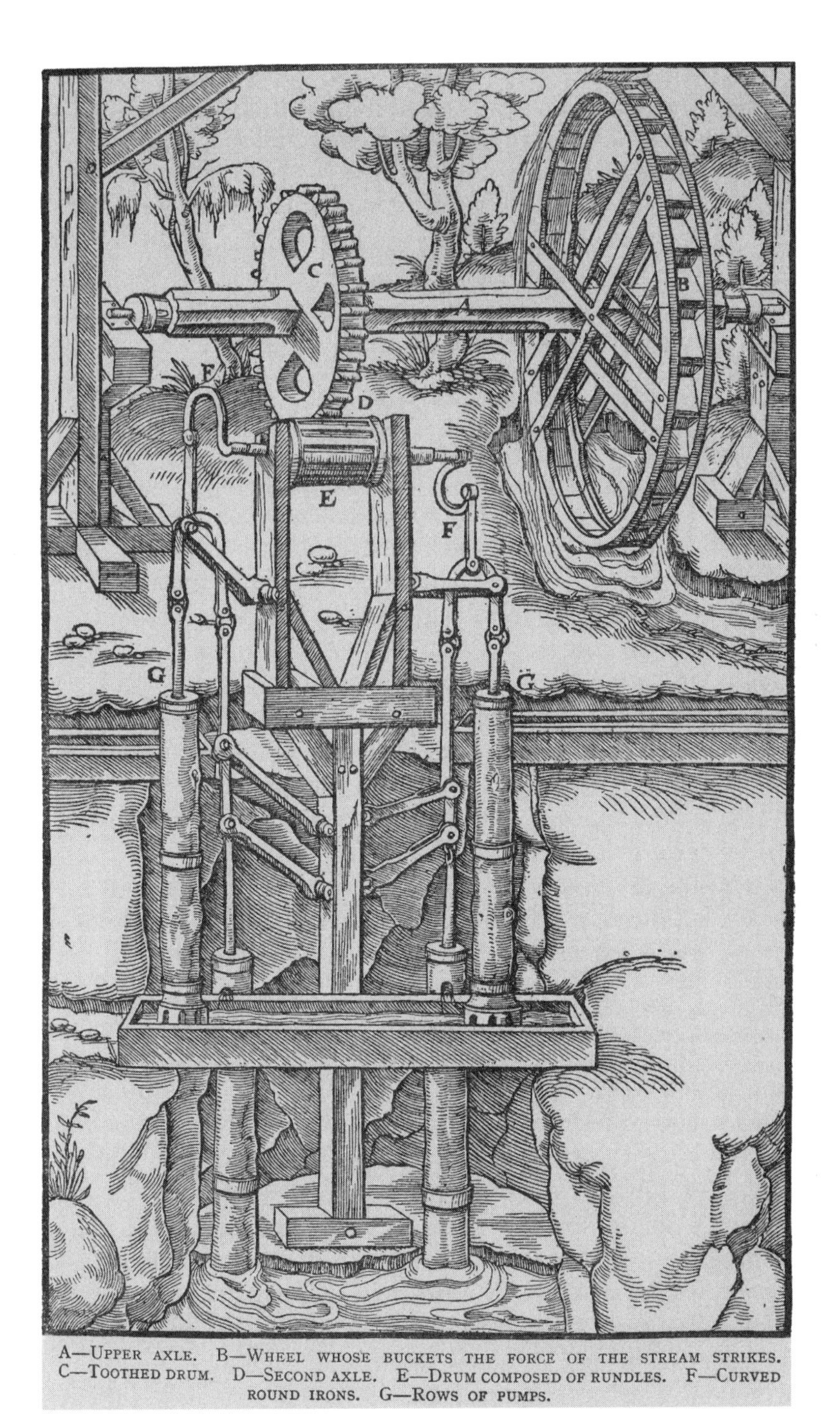

Figure 2.7. Cutaway figure of a two-stage pump for draining water from mines from *De re Metallica*, illustrating the humanists' fascination with mechanical devices. Courtesy of Special Collections, Honnold/Mudd Library.

than from Aristotle's works, in order to undermine scholastic values and to shift natural philosophy in a more applied direction. Then in 1549, he published his own introductory textbook on natural philosophy, *Initia Doctrinae Physicae.* Not by accident, Wittenberg also became a major center for the dissemination of Copernican astronomy because of the practical computational advantages that it claimed to offer to those, including Melancthon who were interested in astrology.

Though Aristotle's aristocratic attitudes were successfully attacked, it proved much more difficult for humanist reformers to abandon many of his fundamental natural philosophical ideas, for those were often deeply imbedded in the astrological, alchemical, and natural historical texts which formed the new focus of humanistically oriented science. In addition, it gradually became clear to Catholic Church figures that the emphasis on the divinity of each and every man that was promoted in connection with Hermetic and other forms of natural magic, was all too compatible with radical Protestantism and the rejection of Church authority. The Council of Trent now declared magical practices—even natural magic—inconsistent with the Catholic faith. Saint Thomas' thirteenth-century fusion of Christian theology and Aristotelian natural philosophy became the official foundation of the new curricula promulgated in such Catholic universities as that at Coimbra in Portugal, and new editions of St. Thomas' commentaries on Aristotelian texts were issued. Thus, the Christian humanist emphases on the divinity of man and on the search for natural knowledge applicable to the improvement of the human condition also produced a conservative backlash within both religion and natural science.

THE LIFE AND WORKS OF PARACELSUS

There was probably no such thing as a "typical" Christian humanist natural scientist, because by the mid-sixteenth century there were a number of different and often mutually antagonistic traditions that grew out of the original emphasis on the divine element in man and the interest in knowledge for the sake of operating in nature. Some groups attacked virtually everything associated with Aristotle and promoted the traditions of applied mathematical knowledge associated with Archimedes. Others sought to revive a pre-scholastic Aristotle and argued strongly against the use of mathematics in the sciences. Among those who did argue in favor of mathematical interpretations of nature, some promoted the Cabala and various forms of number mysticism. Others rejected these uses of mathematics in favor of the Archimedean tradition, while others, including John Dee and Johannes Kepler, were equally at home with the mystical and practical uses of mathematics. Some Christian humanists continued to emphasize the recovery of ancient texts and their translation into elegant Latin. Others pushed for the discovery of new knowledge and the recovery of knowledge embedded within folk and craft traditions, and they urged that such knowledge be communicated in the vernacular so that it was available to much broader segments of the population than it had been in the Middle Ages.

Within this last group, the proponents of alchemy were among the most prolific and insistent. And among alchemists, the medical alchemist widely known as Paracelsus was certainly the most significant. Moreover, his works illustrate the importance of the neo-Platonic worldview that had been revitalized in the Hermetic tradition particularly well. This neo-Platonist, magical approach to the world continued to live as an important element within "modern" science through the seventeenth century, in spite of its strangeness to twentieth- and twenty-first-century readers. Isaac Newton, for instance, was deeply involved with the study of Paracelsan alchemy.

Phillipus Aureolis Theophrastus Bombastus von Hohenheim, who later claimed the title Paracelsus (i.e., greater than Celsus, the Roman medical writer) was born in 1493 just outside of Zurich to a physician and occasional student of alchemy. He studied briefly with a famous Christian cabalist and alchemist, Johannes Trithemius, and then did what many young scholars did: he studied at several universities, ending up at Ferrara, from which he got his medical degree in 1516. By the time that he finished his formal medical education, Paracelsus had already become committed to overthrowing the foundations of traditional academic medicine, which included the writings of the Alexandrian physician and anatomist, Galen of Pergamum (c. 130–c. 201), as well as the *Canon of Medicine* of the great Islamic medical authority, Avicenna (980–1037). His antagonism to medical tradition arose both because its sources were not Christian and because it was ineffective in treating most illnesses. In lieu of ancient written sources of knowledge, Paracelsus placed the Christian magus in direct confrontation with the astral and terrestrial worlds as well as the practical knowledge of ordinary craft practitioners.

Like the Hermetic writings, those of Paracelsus argued that our capacity to understand and act in the natural world came from the stars. Despite the fact that the stars are, in some sense, the source of our natural knowledge and the governors of the natural world, they do not completely govern man because of the divine spark within him:

> In the course of the year, it snows, or rains, hails, and is hot or cold; this is how the heavens constitute the year; and it should be understood that the heavens work in us in a similar way. But we are much stronger than the year, for we can ward off the weather, and seek out the good at the expense of the bad. For we have within ourselves an everlasting summer, which is never without fruit or flowers. . . . Thus we should mobilize our inner powers so that we are not directed by the heavens but by our wisdom. For if we forget this wisdom, so are we like beasts and shall live as reeds in the water and not know from one moment to the next from whence a gust will come and where we will be blown. (Webster 1982, 23)

Thus, although Paracelsus and his followers argued that astral influences are critically important in guiding natural phenomena, they were highly critical of judicial astrology and horoscopic astrology, which purported to be able to predict human events. "The wise man rules over both the sidereal and the elemental, or material body . . . he who imitates the image of God will conquer the stars" (Paracelsus 1951, 155–156).

The key question at this point is how men learn what the stars have to teach us. For Paracelsus, the prospective human magus must first be granted a special divine grace to allow him to recognize natural truths. Then he must seek out knowledge through the constant investigation of nature rather than books. The first way to gain the knowledge that nature has to offer is through experience and experiment. As a consequence, Paracelsus urges the physician to travel and to gain as much experience of plants, animals, and minerals as he can. Then, according to Peter Severinus, one of Paracelsus's earliest and closest followers, one should "purchase coal, build furnaces, watch and operate with the fire without wearying. In this way and no other you will arrive at a knowledge of things and their properties" (Debus 1978, 21).

Sometimes the kind of "experience" Paracelsus advocates is far outside the range of experiences that we now accept as appropriate in the sciences. For example, he advocated recognizing the "signs" that God and the planets placed in objects in order to determine their uses. "Behold the Satyrion root," he wrote, "Is it not shaped like the male privy parts? Accordingly magic discovered it and revealed that it can restore a man's virility and passion"(Boas 1962, 182). More often his experiences reflected an empirical orientation that is easily recognized today.

Rejecting the old humoral explanation for all disease, according to which every disease was the consequence of some imbalance of the four humors (blood, phlegm, black bile, and yellow bile), Paracelsus observed that some diseases seemed to be associated with specific environmental contaminants and that many diseases seemed to affect single organs. In this connection, Paracelsus offered the first description of an occupational disease, the "black lung" disease suffered by many miners, and correctly identified its cause as particulate matter in the air that miners were constantly exposed to underground.

His investigation of specific remedies for specific diseases led him beyond the traditional herbal pharmacopoeia to an extensive investigation of chemical—usually metal-based—medications. He was certainly led to try some compounds as a consequence of macrocosm-microcosm correspondences. For example, because the heart in the microcosm, man, corresponded to gold in the order of metals, Paracelsus expected gold compounds to be effective in curing diseases of the heart. Many critics complained that Paracelsan medicines were, in fact, poisons. But Paracelsus quite correctly argued that dosage makes a huge difference. Some things that are destructive in large amounts can serve as effective medications when ingested in small amounts. Experiment alone, he argued, can determine effective dosages for medicines under various conditions.

Finally, in yet another move to reject scholastic medicine and thumb his nose at academic authority, Paracelsus emphasized that much medical knowledge was to be found among those excluded from power in society. "The physician does not learn everything he must know and master at high colleges alone," he wrote. "From time to time he must consult old women, gypsies, magicians, wayfarers, and all manner of peasant folk and random people, and learn from them; for these have more knowledge about such things than all the high colleges" (Paracelsus 1951, 57).

Though Paracelsus had many friends and clients among both Lutheran and more radical Protestant groups, he considered himself a loyal Catholic until his death in 1541. But he certainly cultivated and achieved a reputation as a social radical and an enemy of corporate authority. When he settled down in Basel in 1527, after about ten years of travel, for example, he insisted upon giving his medical lectures in German rather than in Latin. In addition, he insisted upon making his own medicines, in violation of city regulations that gave the right to compound medicines to the guild of apothecaries. As a consequence, he was forced to leave and take up his wanderings once again.

In 1536, Paracelsus's *Prophecy for the Next Twenty Four Years* appeared. Like many of those influenced by Hermetic ideas, Paracelsus had become fascinated by Joachimite millenarianism and with the interpretation of both biblical prophecies and extraordinary events that seemed to be signs of God's special intervention in the world. *Prophecy* argued that recent celestial events, including a great conjunction (a circumstance in which all of the planets can be seen together within a single zodiacal sign) in 1524 and a spectacular comet in 1531, were signs of the beginning of a period of great turbulence in which ordinary people would be raised up and which would end in the unification of the elect with God.

Though he was careful about identifying specific dates, Paracelsus seemed to suggest that the final assault on the Antichrist was imminent, and this made *Prophecy* an especially popular work among religious and political radicals in England during the period of the civil war in the mid-seventeenth century. Moreover, his popularity as a millenarian prophet generated a wave of interest in Paracelsan medicine and alchemy as well—again, especially among religious radicals (sometimes called enthusiasts). As a consequence, we find major seventeenth-century radical intellectuals studying Paracelsus, writing compilations of craft-based knowledge, as was the case in Charles Webster's *Metallographia: or an History of Metals* (1671) (see Figure 2.8), and promoting educational reforms that would place alchemy and other experimental sciences at the center of higher education.

The identification of Paracelsan alchemy and neo-Platonic natural magic with religious radicalism and with republican political sentiments was certainly not complete, as royalist Sir Elias Ashmole's promotion of natural magic and his alchemical compilation, *Theatrum Chemicum Britannicum* (1652), attest. But the frequent connection between Paracelsan and neo-Platonic views on the one hand, and religious heterodoxy on the other, was important. Robert Boyle's conversion from Paracelsan views to mechanical philosophy was, for example, to a large extent caused by his perception of the unorthodox religious implications of the former.

CHRISTIAN UTOPIAS AND THE INSTITUTIONS FOR MODERN SCIENCE

For the most part, for those who considered themselves followers of Hermes Trismagistus, Paracelsus, or other forms of neo-Platonic

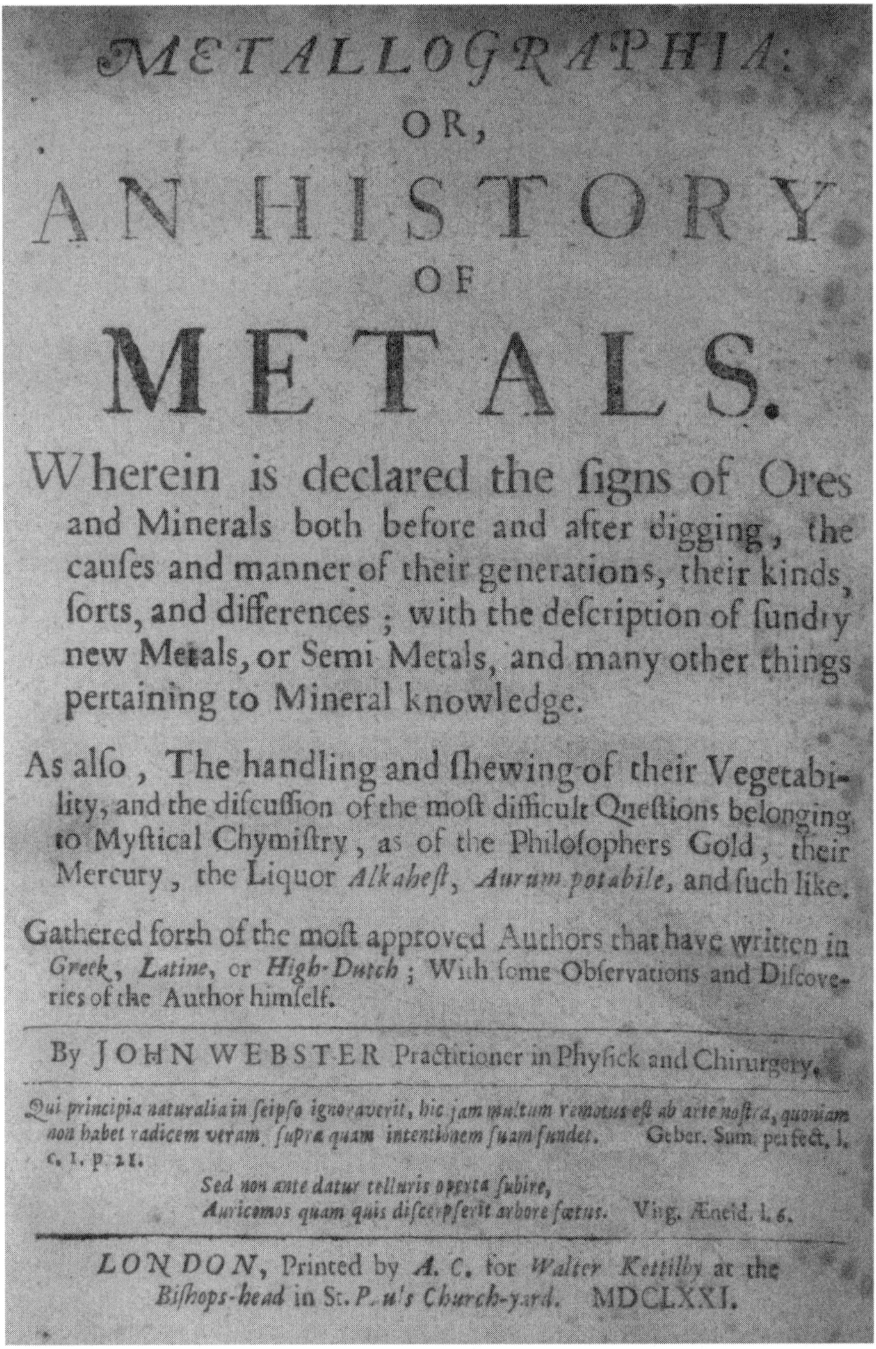

METALLOGRAPHIA:

OR,

AN HISTORY

OF

METALS.

Wherein is declared the ſigns of Ores and Minerals both before and after digging, the cauſes and manner of their generations, their kinds, ſorts, and differences; with the deſcription of ſundry new Metals, or Semi Metals, and many other things pertaining to Mineral knowledge.

As alſo, The handling and ſhewing of their Vegetability, and the diſcuſſion of the moſt difficult Queſtions belonging to Myſtical Chymiſtry, as of the Philoſophers Gold, their Mercury, the Liquor *Alkaheſt*, *Aurum potabile*, and ſuch like.

Gathered forth of the moſt approved Authors that have written in *Greek*, *Latine*, or *High-Dutch*; With ſome Obſervations and Diſcoveries of the Author himſelf.

By JOHN WEBSTER Practitioner in Phyſick and Chirurgery.

Qui principia naturalia in ſeipſo ignoraverit, hic jam multum remotus eſt ab arte noſtra, quoniam non habet radicem veram ſupra quam intentionem ſuam fundet. Geber. Sum. perfect. l. c. 1. p. 21.

Sed non ante datur telluris operta ſubire,
Auricomos quam quis diſcerpſerit arbore fœtus. Virg. Æneid. l. 6.

LONDON, Printed by *A. C.* for *Walter Kettilby* at the *Biſhops-head* in St. *Paul's Church-yard*. MDCLXXI.

Figure 2.8. Title page of John Webster's *Metallographia: or an History of Metals* (1671). This work, with its emphasis on craft traditions, was typical of those produced by radical Protestant intellectuals during and after the English Civil War. Courtesy of Special Collections, Honnold/Mudd Library.

natural magic, the acquisition of natural knowledge was an individual activity that depended to a great degree on the spiritual status of the magus. A very different viewpoint began to develop within the natural magic tradition, however. Among the most widely distributed and admired of the books of secrets published in the sixteenth century was Giralomo Ruscelli's *Segerti Nuovi,* published first in 1555. This book, which went though more than seventy editions in several languages before the end of the seventeenth century, purported to reveal the results of experiments done by a secret society, the *Accedemia segreta,* at Naples. Only after a "secret," or recipe, was shown to be reliable in multiple trials within the group was it publicly disseminated.

The notion implicit in this work—that is, that experimental knowledge should be the product of a collaborative enterprise and that collectively produced and tested knowledge gained credibility over privately produced knowledge—was embodied in a small and short-lived local society established by the natural magician Giambattista Della Porta at Naples in 1560 under the title *Accedemia dei segreti*. More importantly for the long run, it was strongly promoted in a group of extremely popular Christian humanist utopias, all written within a few years of one another at the beginning of the seventeenth century. Johann Andreae's *Christianopolis* (1619), Tommaso Campanella's *City of the Sun* (1623), and Francis Bacon's *New Atlantis* (1626) all offered visions of a new and perfected society. At the heart of each lay a set of institutions for the collective discovery and dissemination of natural knowledge aimed at improving human welfare. Collectively these works seem to have both stimulated widespread enthusiasm for experimental natural philosophy and offered models emulated by the founders of such major new institutions for the pursuit of the sciences as the Royal Society of London for Improving of Natural Knowledge.

Andreae was born at Herrenberg in what is now southern Germany in 1586. His father, a Luthern pastor, died when he was very young and his mother became an apothecary and boarded medical students; so Johann became interested in drugs and was exposed to both traditional and Paracelsan medicine as a child. In 1603, he received his bachelor's degree and began to study theology; but he continued scientific studies on the side, studying with Michael Maestlin, an outstanding Copernican who mentored Johannes Kepler, with whom Andreae carried on a lifelong correspondence. Through Maestlin, Andreae also met Christopher Besold, a local student of Hermetic writings and alchemy, and Andreae became fascinated with this literature.

Soon he was on his way to becoming one of the most prolific authors and successful social reformers of his day.

In 1619, Andreae published *Christianopolis,* modeled on Thomas More's *Utopia.* As in More's work, the central figure is shipwrecked and washes up on an island that is the site of a perfect society. He is interrogated by a delegation from the nearby city to determine whether he is a pious Christian and whether he has made progress in "the observation of the heavens and the earth, in the close examination of nature, [and] in instruments of the arts" (Andreae 1916, 148). When he says yes, he is given a group of guides to show him Christianopolis. Like Utopia, Christianopolis is a communistic society in which goods are held in common. Its governmental structure is that of a theocracy, and its theology is orthodox Lutheran. But what attracts the reader's attention above all else is the relationships among scientific research, education, and economic activity.

The first stop in the tour is the metallurgical workshop:

> [Here] Everything that the earth contains in her bowels is subjected to the laws and instruments of science. Here men are not driven to work with which they are unfamiliar like pack animals to their task, but they have been trained long before in accurate knowledge of scientific matters, and feel their delight in the inner part of nature. If a person does not here listen to reason and look into the most minute elements of the macrocosm, they think that nothing has been accomplished. Unless you analyse matter by experiment, unless you improve the deficiencies of knowledge by more capable instruments, you are worthless. . . . To be brief, *here is practical science.* (Andreae 1916, 154–155; emphasis mine)

Proceeding to the center of the city, we find among the buildings devoted to government, religion, and education, a series of teaching laboratories, anatomical theaters, natural history museums, and collections of mathematical instruments. In each, men learn to "assist the struggles of nature" (Andreae 1916, 199). After describing the Christianopolitan's education in mathematics and natural philosophy, which combines both knowledge designed to increase one's appreciation of the universe as well as knowledge designed to increase one's ability to act on it, Andreae concludes with the following remarkable statement:

> It is . . . man's *duty,* now that he has all creatures for his use, to give thanks to God himself in place of them all; that is, he should offer to God as much

obedience as he observes in His creatures. Then, *he will never look upon this earth without praise to God or advantage to himself. . . . Blessed are they who use the world and are not used by it.* (1916, 231–232; emphasis mine)

In Andreae's hands and those of his near contemporaries, including Francis Bacon, the notion of "using" nature was always governed by religious priorities. Thus, for example, Bacon writes in his *Great Instauration*:

I would address one general admonition to all, that they consider what are the true ends of knowledge, and that they seek it not either for pleasure of the mind, or for contention, or for superiority to others, or for profit, or for fame, or power, or any of these inferior things, but for the benefit and use of life; and that they perfect and govern it in charity. For it was from lust of power that the Angels fell, from lust of knowledge that man fell, but of charity there can be no excess, neither did angel or man come in danger by it. (1937, 251)

Andreae attempted to do what he preached. On a larger scale, and directly related to the vision of social betterment through applied science presented in *Christianopolis*, he sought to establish a *Societas Christiana* of learned men who would band together to improve the lives of everyone. In particular, he sought to bring together physicians, anatomists, mathematicians, mechanics, and chemists. Though that society never seems to have materialized, Andreae's ideas were taken up in England by a group of intellectuals surrounding Samuel Hartlib (who had Andreae's ideas translated as *A Modell of a Christian Society*). This group did try to establish a collaborative effort of scientific intellectuals under the government's Office of Address during the 1650s, and they were among the groups that established the Royal Society of London in 1662 after the restoration of the monarchy.

Andreae's ideas, and Hartlib's attempts at implementing those regarding the promotion and use of natural knowledge, undoubtedly seemed dangerous to some members of society because of their close association with socially radical ideas. Without withdrawing from the emphasis on using natural knowledge for general human betterment, Francis Bacon managed to present the core ideas of Christian humanist natural knowledge in a manner that was more palatable across the social and political spectrum. Bacon's views were widely admired by Puritan radicals such as Hartlib, but Bacon's experience as a (briefly) powerful member of the royal court—he served as Lord

Chancellor under James I—allowed him to formulate his ideas in a less threatening way. When Bacon offered his utopian vision in *The New Atlantis*, he offered an even more elaborate network of laboratories, anatomy theaters, and processes for seeking practical knowledge from around the world than Andreae had offered, but his island of Bensalem was governed by a hereditary monarchy, and social stratification was maintained and marked by distinctions of dress and behavior. So it was primarily to Salomon's House, Bacon's projected society for discovering new practical knowledge, that men from across the British religious and political spectrum pointed when they sought support to institute the Royal Society of London.

The notions of spiritual perfection and of material improvement have gradually become disassociated from one another in the modern world. All too often knowledge does seem to be pursued primarily for profit or to establish military or economic superiority over others. But it is worth remembering that at the origins of the modern scientific emphases on experiment and gaining natural knowledge for application in the world lay a serious religious impulse associated with Christian humanism.

Chapter 3

Science and Catholicism in the Scientific Revolution, 1550–1770

Since A. D. White's *History of the Warfare of Science with Theology in Christendom* appeared in 1896, virtually every discussion of the relationship between Catholicism and science in early modern Europe has begun with the Galileo trial. Furthermore, almost all have insisted that the trial of Galileo and the doctrines that underlay it put a severe damper on science in Catholic countries and turned them into intellectual backwaters. With respect to astronomy in particular, as late as 1986 a distinguished historian of science could still claim that in the aftermath of the events of early 1633, "Cosmological discussion ceased except among the Jesuits, and astronomy, with the single exception of Giovanni Borelli was reduced to the making and using of telescopes" (Ashworth 1986, 153). Thirteen years later, however, the non-Catholic historian of science, John Heilbron, argued that "The Roman Catholic Church gave more financial support to the study of astronomy for over six centuries, from the recovery of ancient learning during the late middle ages into the Enlightenment, than any other, and, probably, all other, institutions"(Heilbron 1999, 3). What happened? Was the Roman Catholic Church the most effective institutional agent for the repression of early modern intellectual inquiry—especially scientific inquiry—or was it the most generous sponsor of scientific activity in early modern Europe?

The answer to this question seems to be a bit of both. It is certainly true that the Counter-Reformation Church, starting just before the Council of Trent (1545–1563), became deeply concerned with doctrinal conformity. Pope Paul III established the Holy Office of the Inquisition in the Vatican in 1542 and the Index of Prohibited Books in 1543 to ensure conformity to accepted doctrine. Many natural philosophers found themselves involved with these two institutions over the next 200 years, although those interested in natural magic and alchemy were far more stringently dealt with than astronomers. It was, for example, because of his advocacy of Hermetic magic and his claim that Moses and Christ were magi and not for any astronomical views that Giordano Bruno was condemned by the Holy Office of the Inquisition and burned to death on February 17, 1600.

In 1539, just before establishing these policing institutions, Pope Paul III established the Society of Jesus (the Jesuits), the only monastic order to report directly to the papacy, to ensure the education of Catholics in proper doctrine and to undertake the conversion of non-Catholics to Catholicism. The Jesuits became the chief teachers of Europe—at least Continental Europe—as well as the chief agents for the spread of European knowledge throughout the rest of the globe, for the next two-and-a-quarter centuries (until just after they were expelled from France in 1768). For reasons to be considered later in this chapter, the Jesuit order found ways to promote scientific activity that were technically within the bounds of doctrinal conformity, but which certainly stretched the boundaries of officially allowed discourse. After 1633, for example, they continued to teach elements of Copernican astronomy, though always as a hypothesis. Indeed, in the mathematical sciences, in astronomy, and in experimental natural philosophy, Jesuits were among the most prolific scientists in Europe; they stimulated the scientific activities of the vast majority of secular scientists on the Continent; and their order almost certainly did outspend any and probably all other institutions in Europe on astronomical observatories and experimental equipment through the seventeenth century and the first half of the eighteenth century.

Few Jesuits were involved in the development of the mechanical philosophy, the major seventeenth-century innovation in scientific theory. Nonetheless, it was largely through Catholic scholars, including the lay Catholic René Descartes (1596–1650), the Minim friar Marin Mersenne (1588–1648), and the secular priest Pierre Gassendi (1592–1655), that this major alternative to both Aristotelian and neo-Platonic philosophy was promoted during the first half of the seventeenth century.

In this chapter we will explore some of the reasons behind the Trentine decision to link Catholic doctrine to Aristotelian natural philosophy—especially as interpreted by Saint Thomas Aquinas. We will look at the very different ways in which the Dominican order and the Jesuit order chose to interpret and apply Trentine doctrines, especially as they related to the sciences. We will very briefly revisit the Galileo case in terms of the conflict between Dominican and Jesuit understandings of appropriate scientific method. We will follow the transformation of an order whose focus on teaching official doctrinal conformity changed into one whose major activities included the teaching of the exact sciences and promotion of experimental natural philosophy. We will consider both the strengths and the weaknesses of Jesuit science as they relate to religious doctrines. Finally, we will consider a few Catholic intellectuals outside of the two most powerful orders and explore the relationship between their religious commitments and their adoption of various versions of the mechanical philosophy.

SCIENCE AND THE COUNCIL OF TRENT

By 1540, substantial portions of northern Europe and all of England had formally parted from the unified Roman Catholic Church, and Pope Paul III was deeply concerned with ensuring both that further defection be stopped and that rebellious groups be brought back into the Church. In order to deal with a variety of issues related to a counterreformation, he called a church council to be held at Trent in Italy beginning in 1545. The complexity of the issues facing the Church might be suggested by the fact that the council met off and on through 1563, and even then, as we shall see, some of the issues it raised were never completely resolved.

One of the important aims of the council was to set forth Catholic doctrine and make clear the differences between Catholic and Protestant beliefs and practices. Within this issue, the interpretation of what happens in the sacrament of the Eucharist became particularly important with respect to Catholic natural philosophy, as did the doctrines of predestination and human free will. Some of the attention of the council was also absorbed by issues of how to ensure conformity to established church doctrine without being so restrictive that good Catholics were driven away, and by how to bring Protestants back to the Church.

The sacrament of the Eucharist constitutes the central ritual of the

Catholic mass. In this sacrament, wafers of bread and a goblet of wine are prepared (consecrated) by a priest and, according to Catholic doctrine, they are converted into the body and blood of Christ, which congregants then eat and drink to place them in a state of union with Christ. This alone can give meaning to the notion that Christ died for the congregants' sins. According to Protestant doctrines, the wafers and wine are not literally converted into Christ's flesh and blood; they merely symbolize or stand as signs for them. Thus, the entire ritual merely symbolizes a union with Christ; it does not constitute a real union. On such an important point the Council of Trent felt it critically important to emphasize the Catholic position that a real change occurred so that the body and blood of Christ are present "in very truth," and not "merely as in a figure or sign."

The major scriptural authority for the real presence came from the Gospel of John, chapter 6, verses 51–57, in which Christ insists that "the bread that I give is my own flesh; I give it for the life of the world," and "in very truth I tell you, unless you eat of the flesh of the Son of Man and drink of his blood you can have no life in you. Whoever eats my flesh and drinks my blood possesses eternal life, and I will raise him up on the last day." Yet just a few verses later, Christ says, "The spirit alone gives life, the flesh is of no avail," as if he were acknowledging that the flesh and blood mentioned earlier were only signs of a spiritual transformation in men. So these biblical passages alone were not sufficient to establish Catholic doctrine. Similarly, while many of the early church fathers, including Ambrose, had supported the claim of a real presence in the Eucharist, others, including Augustine, had argued for a symbolic presence.

There was, however, one source beloved of the Dominicans—in part because he was one of them, and in part because his stock as a great theologian had been growing—who insisted that a real "transubstantiation" (i.e., change of substance) took place. Furthermore, he purported to be able to offer a compelling explanation of how it could be so. That man was Saint Thomas Aquinas, at once one of the major medieval theologians as well as one of the most prolific European commentators on Aristotelian natural philosophy. The council drew from Thomas Aquinas' *Summa Theologica, Part III* (Questions 73–77), to affirm and explain the real presence of Christ in what appeared to be bread and wine in the Eucharist. And it did so in terms of concepts drawn from Aristotelian natural philosophy, because that was the conceptual structure within which Thomas's theology had been developed.

At the height of the thirteenth-century conflicts over whether Aristotelian philosophy could be brought into line with Christian theology, Thomas had carved out a middle ground between those admirers of Aristotle who demanded a wholesale revision of theology and those conservative theologians who demanded a virtual abandonment of Aristotelian philosophy. Thomas insisted that with respect to those topics to which natural philosophy was properly applied, there can be no conflict between reason—represented by Aristotelian philosophy—and revelation—represented in Scripture. When an apparent conflict occurs, it has to be because we erroneously interpret the claims or erroneously judge the applicability of one or the other. There might, however, be supernatural phenomena for which revelation alone can account.

According to Thomas, Aristotle correctly claimed that there can be no natural reason to support the belief that the world has not existed forever. But revelation makes us aware that God created the world out of nothing, at some point in time. There is an apparent conflict between theology and natural philosophy, one arguing for an infinite age for the universe, the other for a finite age. But, according to St. Thomas, this conflict disappears when we realize that philosophy only denies the possibility of a natural reason for a particular beginning of the world; it does not deny the possibility of a supernatural reason. Similarly, natural reason offers no possibility that the substances of bread and water might be really converted into flesh and blood without any change in the sight, texture, flavor, and smell. Nor could it account for Christ's simultaneous presence in heaven and in all of the wafers and wine being blessed throughout the world at any time. Yet, Saint Thomas' theology, borrowing from the language of Aristotelian natural philosophy, was able to account for both.

Thomas begins from three basic Aristotelian notions.

1. Every "substance,"—that is, every particular thing—is composed of matter and form.
2. We come to know substances only from their "accidents"—sensible qualities such as size, shape, taste, and texture, and their relationships to other substances such as place, kinship, and temporal proximity—which inhere in them.

3. One substance can be "transformed" into another by keeping the matter the same and changing the form. This is how Aristotle understands the growth of a tadpole into a frog, for example. It is a kind of transformation. In nature, every transformation is limited by the fact that particular substances have within them only the potential to become certain things rather than others. A tadpole, for example, can naturally be transformed into a frog but not into an oak tree, just as an acorn can be transformed into an oak but not into a frog.

Now we come to the question of what occurs during the consecration of the bread in the ritual of the Eucharist. Thomas seems to back into this issue by asking first whether the substances of bread and wine remain after the consecration. To provide a sense of Thomas' language, here is the central portion of his answer:

> a thing cannot be in any place where it was not previously, except by change of place, or by the conversion of another thing into itself; just as fire begins anew to be in some house either because it was carried there, or because it was generated there. Now it is evident that Christ's body does not begin to be present in this sacrament by local motion. First of all, because it would follow that it would cease to be in heaven: for what is moved locally does not come anew to some place unless it quit the former one. Secondly, because every body moved locally passes through all intermediary spaces, which cannot be said here. Thirdly, because it is not possible for one movement of the same body moved locally to be terminated in different places at the one time, whereas the body of Christ under this sacrament begins at the one time to be in several places. Consequently it remains that Christ's body cannot begin to be anew in this sacrament except by the change of the substance of the bread into itself. But what is changed into another thing, no longer remains after such change. Hence the conclusion is that, saving the truth of this sacrament, the substance of the bread cannot remain after the consecration. (Aquinas 1947, Question 75, article 2)

After determining that the bread cannot remain after the consecration Thomas asks how the bread is turned into the body of Christ. At this point, he argues that the change is unnatural, for bread does not have a natural potential to be converted instantaneously into flesh. The change must be accomplished by God's supernatural agency, and as a consequence, this change deserves a special name of its own:

> He [God] can work not only formal conversion, so that diverse forms succeed one another in the same substance; but also the change of all being, so that,

to wit, the whole substance of one thing be changed into the whole substance of another. And this is done by Divine power in this sacrament; for the whole substance of the bread is changed into the whole substance of Christ's body, and the whole substance of the wine into the whole substance of Christ's blood. Hence this is not a formal, but a substantial conversion; nor is it a kind of natural movement: but with a name of its own, it can be called "transubstantiation." (Aquinas 1947, Question 75, article 4)

Among the problems that this account leaves is how we continue to see and taste the characteristic "accidents" associated with bread and wine after they have been converted into the body and blood of Christ. Here again Thomas argues for the suspension of the natural links between substances and accidents made possible by God's omnipotence:

The accidents continue in this sacrament without a substance. This can be done by Divine power; for since an effect depends more on the first cause than the second, God, who is the first cause of both substance and accident, can by his unlimited power preserve an accident in existence when the substance is withdrawn whereby it was preserved in existence as by its proper cause, just as without natural causes He can produce other effects of natural causes, even as he formed a human body in the Virgin's womb, "without the seed of man." (1947, Question 77, article 1)

Thomas' version of Aristotelianism held a special appeal for virtually all parties at Trent because it provided a framework for understanding the Catholic insistence upon transubstantiation during Eucharist; but his attack on astrology held a special appeal for the Jesuits in connection with a key doctrinal issue that was never fully resolved at Trent. That issue concerned the Calvinist Protestant notion of predestination and the relationship between grace and free will. The basic question at issue was how it could both be true that God determines in advance everything that will happen in the universe and that human beings are free to choose to act morally or immorally—which is what it means to have free will.

The Dominicans, who emphasized God's omnipotence, claimed to follow Thomas' theology most closely on this issue. They argued that predestination means that God wills man's free will to cooperate with his grace. That is, God causes humans to choose as they do, so that some specified persons choose the path to salvation while others do not. In this sense, "the relationship between God's grace and man's will resembles that between a stick and its mover. By definition, no stick will move until moved by a mover" (Feldhay 1987, 205).

For the Jesuits, who wanted to offer every believer the hope that one's effort to do good works might warrant God's grace, such a doctrine made the concept of free will meaningless. In order to maintain human free will and God's omniscience, the Jesuits argued that God's foreknowledge and his will should be separated. He has absolute and certain knowledge of the future acts of all humans; and in some sense, this knowledge guides his will in deciding who will be saved. Both God's knowledge and his will precede the actual acts of the human; thus, the acts may be said to be predestined by God. But human free will is preserved through a kind of time-reversed causality. What humans will do in the future causes God's preceding gift of grace. Such an argument demanded that traditional Aristotelian and Thomist notions of logic be revised; for according to medieval logic, there could be no absolute knowledge of future contingent acts or of what the Jesuits interpreted as "hypothetical" (not realized) acts.

One consequence of this argument was that the Jesuits revised the way in which "hypothetical" entities, including (1) the future acts of humans and (2) the very existence of mathematical entities (which also had the character of not being physically realized), were understood. Previously, "hypothetical" implied "false," so consequently hypotheticals could not be the subject of any true science. Now it became possible to claim absolute, certain knowledge of hypothetical entities even though they did not "exist" in the usual physical sense. Only God, of course, could have knowledge of future human acts; but even humans could have knowledge of mathematical entities, such as those posited in astronomical systems. In modifying the Thomist understanding of Aristotelian logic, the Jesuits argued that while Thomas provided the most reliable guide on most issues, when some issue could not be satisfactorily resolved using Thomist philosophy, it was permissible to think outside of that system.

If human choices were governed by the stars or by any other occult forces, human free will would, of course, be illusory; so the Jesuits turned away from the neo-Platonic and Hermetic doctrines that focused on astral influences, instead drawing heavily from Aquinas' critique of astrology for this purpose. Thus, though Dominicans and Jesuits could not agree on how to understand predestination and free will, they could and did agree that any neo-Platonic or Hermetic understanding of the world was unacceptable—because it was inconsistent with Saint Thomas' Aristotelian philosophy for the one and because it was inconsistent with the notion of the absolute freedom of the human will for the other. All neo-Platonic philosophies that viewed humans as magi, manipulating the world through hidden forces that were not understandable through the "reason" associated with Aristotelian doctrines, were additionally suspect because Thomism depicted magic as converse with evil spirits. No matter how much those who understood themselves as natural magicians insisted that their magic was "natural" rather than "demonic," both Jesuits and Dominicans insisted that since all natural features of the world were to be comprehended within Aristotelian categories, any hidden or occult powers must be demonic.

So far, if we consider the consequences of the Council of Trent on the pursuit of the natural sciences, we can say that widely shared doctrine strongly discouraged the pursuit of those sciences linked closely with Hermeticism and neo-Platonism. Judicial astrology, Cabala, and even Paracelsan alchemy thus found very little support among Catholic scholars during the 200 years following Trent. On the other hand, Thomas' version of Aristotelian natural philosophy was strongly promoted—very rigidly by the Dominicans, and more loosely by the Jesuits. Finally, on certain issues connected with the theory of what kind of knowledge can be attained, the Jesuits argued for the possibility of certain knowledge of hypothetical entities, including the subjects of mathematical astronomy while the Dominicans insisted that true sciences can only have as their subjects entites that actually exist. For this reason, any claim to scientific knowledge of astronomical entities necessarily implied their existence.

The Dominicans, following Thomas, did believe that true knowledge about mathematical astronomy could be achieved, but only if the assumptions it used were derived from the more general, or "superior" science of natural philosophy. Thus, in 1546, the Dominican Giovanni Tolosani argued that Copernicus had been led into error because his system was not grounded in an appropriate causal explanation for why the earth and other planets should move around the sun. Neither Copernicus in 1543, nor Galileo in 1632, was able to offer a physical explanation that could support the mathematical astronomy that they believed established the truth about the heavenly motions.

For the Dominicans, any astronomical hypothesis not grounded in physics was considered false. In the Galileo case, then, either Galileo was proposing the Copernican system as a mere hypothesis, in which case it was false according to this standard; or he was claiming that it was true, in which case he was wrong because the Copernican system violated the dominant Aristotelian natural philosophy. Indeed, it would not be until Isaac Newton published his *Mathematical Principles of Natural Philosophy* in 1687 that a credible physical cause could be offered for a sun-centered theory of planetary motion, and that theory was a revision of the Copernican system developed by Johannes Kepler.

JESUIT SCIENCE

In connection with their disagreements at Trent over issues such as free will, both the Dominicans and the Jesuits formulated or reformulated the goals and doctrines of their own orders, which had a major impact on Catholic science. In 1566, the Dominican order issued the first of a series of new rules which added to the original Augustinian rules chosen by the order. These rules reaffirmed the order's vocation in terms of the traditional Aristotelian life of contemplation and insisted that all studies should be directed at the contemplation of divine things. Secular learning was allowed, but there was an insistence that it not be allowed to become a goal in itself. Any time that secular learning seemed to challenge traditional doctrine it was almost automatically attacked. The major consequence of this emphasis was that Dominicans played a minimal positive role in the sciences after 1563.

The Jesuits, on the other hand, came from a more activist and humanist orientation, rejecting the monastic life for one of learning and service. In his original *Constitutions* for the society, Jesuit founder Ignatius of Loyola insisted:

> In the midst of actions and studies, the mind can be lifted to God; and by means of this directing everything to the divine service, everything is prayer. . . . The distracting occupations undertaken for His greater service, in conformity with His divine will . . . can be, not only the equivalent of the union and recollection of uninterrupted contemplation, *but even more acceptable, proceeding as they do from a more active and vigorous charity.* (1970, 184, italics mine)

Jesuits might find themselves assigned to duties in secular courts, to missions throughout the world, or, increasingly often over time, to teaching in one of the many Jesuit colleges or universities in Europe.

At first, it was understood that "studies" meant those directly related to contemplation of the Divine; but by the 1560s, studies directed at any form of truth, sacred or secular, had become a legitimate path to God for the Jesuit. In 1563, a new set of rules argued for two quite different paths:

> Let everybody know that the Society has two means by which it strives for this end: the one is a certain force, spiritual and divine, which is acquired through the sacraments, prayer, and the religious exercise of all virtues, and which is warranted by the special grace of God; and the other force is placed in the faculty which is ordinarily found through studies. (Feldhay 1987, 200)

Given their emphasis on study, the Jesuits allowed substantial latitude to members regarding what they could study and teach. For example, even though Jesuits were committed in a general way to Saint Thomas' interpretation of Aristotle, they allowed members to lecture on the commentaries of the Muslim scholar Averroes on Aristotle, which interpreted him very differently. Moreover, in spite of the order's opposition to magic in general, many Jesuits, including two of the most widely respected Jesuit scholars of the seventeenth century, Gaspar Schott and Athanasius Kircher (see Figure 3.1), culled the book of secrets tradition for descriptions of the hidden virtues of natural objects and promoted the mathematical elements of natural magic.

By the last third of the seventeenth century, toleration even for such anti-Aristotelian philosophies as that of René Descartes became common; so when a local prelate complained that the Jesuit mathematician Honore Fabri had gone beyond what was appropriate in praising Galileo, Descartes, and Copernicus, the Secretary-General of the order wrote to him, arguing that, "one can indulge a man of such parts (i.e., great abilities)" (Heilbron 1982, 106) And in spite of the inquisitorial treatment of Galileo, many Jesuit astronomers taught the Copernican system and admitted its superiority to geocentric systems for purpose of many calculations, although they did consider it only one of a number of useful hypotheses. Jesuit scientific treatises occasionally ended up on the Dominican-dominated index of prohibited works, but such an event rarely interfered with the career of a Jesuit scholar.

The Society of Jesus was also the first early modern institution to establish research positions separated from teaching or court service. Beginning as early as 1565, positions as "Scriptors," free of all obliga-

Figure 3.1. Title page of Athanasius Kircher's *Mundus Subterraneus*. Though doctrines promulgated at the Council of Trent opposed study of the magical and occult arts, Jesuit scholars such as Kircher were given great latitude and produced some of the most complete compendia of natural knowledge, including occult science. Courtesy of Special Collections, Honnold/Mudd Library.

tions other than scholarship and lasting from two to six years, were established for distinguished scholars. Many, if not most, of those positions were held by mathematical astronomers such as Christopher Clavius and Giambattista Riccioli, mathematicians such as Jacques Grandami and Honore Fabri, and experimental natural philosophers such as Noel Regnault. Furthermore, the society and wealthy friends of the society supported both the research and teaching of their scientific scholars by building twenty-five observatories by the end of the seventeenth century and by establishing collections of scientific instruments and natural history artifacts that were the envy of all non-Jesuit educators.

One consequence of this institutional support was that Jesuit contributions to certain fields of early modern science were vastly disproportionate to their numbers. Between 1600 and 1773, some 1,600 different Jesuits contributed nearly 6,000 original scientific works. In the one field for which a detailed analysis of all work done between 1600 and 1789 has been attempted—experimental studies of electrical phenomena—more than 15 percent of all authors were Jesuits and their contributions constituted closer to 30 percent of all published works (derived from data in Heilbron 1982). Initially, many Jesuit scientific works involved the humanistic desire to go beyond imperfect medieval translations of classical works to the originals. Thus, a massive project of the Jesuits at Coimbra in what is now Portugal involved publishing all of the works of Aristotle in the original Greek, with new Latin translations on facing pages, then providing new commentaries, often disagreeing with Saint Thomas. Beginning around 1640, however, it became increasingly clear that Aristotelian natural philosophy was being superseded. At that point, works in applied mathematics, including mathematical astronomy, mechanics, and optics, in observational astronomy, and in experimental natural philosophy overtook and passed Aristotelian-based works.

Jesuit scientific publications reflected the emphasis placed on mathematics, mathematical physics and astronomy, and experimental natural philosophy within the curriculum of Jesuit higher education. This curriculum was established in the *Ratio Studiorum*, which took its final shape in 1599 after going through numerous refinements during the 1580s and 90s, and it provided the framework for Jesuit education through the eighteenth century. Mathematics and natural philosophy dominated the culminating two years of the eight-year curriculum for several reasons.

During the late sixteenth century, the Church had a special need for competent mathematicians and astronomers to address calendar reform. According to church doctrine, Easter was to be celebrated on the first Sunday after the first new moon after the vernal equinox (the day in spring when day and night are of equal length in the Northern Hemisphere). Various methods for computing the date of Easter in the Julian calendar had been proposed throughout antiquity and the early Middle Ages, and a common method had been settled on in 669 C.E. This method, explained by the English author Bede in *De Temporum Ratione* in 725, became the foundation for Easter calculations throughout the Middle Ages. But the actual dating of Easter depends critically on four conditions:

1. a civil calendar in which the length of each year can be predetermined such that the vernal equinox always occurs on nearly the same date
2. an accurate knowledge of the date on which the vernal equinox occurred at some point in time
3. an accurate value of the average month length
4. an accurate value for the average length of a solar year.

The last three of these conditions depend on having precise astronomical observations available, while the first depends on calculations based on the last three.

By the mid-fourteenth century, calculations using Bede's rules placed the celebration of Easter as much as a full month away from the actual date of the first Sunday after the first new moon after the vernal equinox. Recommendations for calendar reform were made at the Council of Constance in 1415, the Council of Basel in 1434, and the Lateran Council scheduled for 1511, but experts could not agree. The Council of Trent referred the problem to Pope Gregory XIII and his advisors. In 1582, the Gregorian Calendar reform was established throughout Roman Catholic Europe. It dropped ten days to bring the vernal equinox to March 21, changed the year length by omitting three leap years in every 400 years, and adjusted the average month length by dropping one day in the lunar cycle in every 312.5 years. This seemed to work. But the Jesuit astronomer Christopher Clavius and others insisted that further modifications might be called for if better values for the length of the solar year and the month could be established. As a consequence, devices for the very accurate observation of the dates on which the sun passed through well-defined points in its path were established in a number of cathedrals, and training in mathematical astronomy was encouraged. (Anyone interested in a more detailed account of the mathematical and observational details of calendar reform should consult Heilbron 1999.)

Mathematics was also emphasized because as the Jesuits increasingly sought to educate not only clerics but also the lay elite, more and more prospective students sought mathematical knowledge to help train them for navigation, surveying, and architecture—all subjects that were important for high-status careers as what we would now call military and civil engineers. Some cities, including Avignon, allowed the Jesuits to establish a college only on the condition that they provide a mathematician and mathematical training to their students. (For a time into the early seventeenth century, there were not yet enough trained Jesuit mathematicians to implement the entire curriculum established in the *Ratio Studiorum*, so not every college had a mathematician.)

Finally, it is very likely that the Jesuits emphasized mathematical approaches to such topics as astronomy, mechanics, and optics, because it allowed them to teach about these subjects "hypothetically" without committing to any particular theories regarding the causes of the phenomena being studied. Thus, they could teach Copernican, Tychonic, Keplerian, Cartesian, and—eventually—Newtonian sciences without danger of violating their doctrinal commitments. It is also likely that the Jesuit emphasis on experimental work in natural philosophy was a consequence of the same pressure. One could experiment on almost any topic without committing to a particular causal explanation of the phenomenon being studied. Ironically, the very Catholic doctrines that are often seen as part of the warfare of Catholicism against science helped indirectly to establish practices that are today seen as typical of modern science.

Perhaps the most important and positive consequence of the Jesuit emphases on mathematical and empirical sciences was that they produced comprehensive textbooks and reference works that were used by Protestant as well as Catholic scientists and laypersons. Giovanni Riccioli's *Almagestum Novem* of 1651 provided much of the astronomical data used by such English Protestant scientists as John Flamsteed and Isaac Newton, for example. Athanasius Kircher's *Mundus Subterraneus in XII Libros Digestus* of 1665 provided much of the science known in the New World throughout the seventeenth century. And Gaston Pardies' *Eléments de Géometrie* of 1681 went through at least eight English editions as *Short, but yet Plaine Elements of Geometry* by 1746. On the other hand, the Jesuits' hesitancy to commit to any single causal explanatory philosophy, such as the mechanical philosophy which constituted one of the major theoretical advances of the seventeenth century, gradually undermined their authority among non-Catholics.

In 1751, an edict of the Secretary-General finally released Jesuit scientists from their commitment to Aristotle. But even then, it demanded acceptance of those specific elements of Aristotelian philosophy which underlay the Thomist accounts of the finite age of the earth and of transubstantiation—that is, the matter/form dichotomy, the possibility of creation out of nothing, and the existence of some absolute independent accidents (Heilbron 1982, 104).

CATHOLICS AND THE MECHANICAL PHILOSOPHY: MERSENNE, DESCARTES, AND GASSENDI

Beginning in the early seventeenth century, a new method of explaining natural phenomena competed with the Aristotelian and neo-Platonic philosophies that had dominated the Renaissance. Scholars began with increasing frequency to liken natural entities to the artificial creations of mechanics, architects, artisans, and what we call engineers, all of which were becoming increasingly important in the dynamic commercial culture of urban Europe. The mechanical philosophy undoubtedly owed some of its appeal to the growing familiarity that people in the seventeenth century had with fabricated and engineered objects; but many of those who adopted it also found religious reasons. In 1605, for example, Johannes Kepler abandoned his neo-Platonic understanding of the world as a gigantic organism in favor of likening it to a clock, and when he did so, he justified it in religious terms that would be repeated by many of those newly converted to the mechanical philosophy. Writing to a close friend, Kepler said:

> I am now much engaged in investigating physical causes. My goal is to show that the celestial machine is not the likeness of the divine being, but is the likeness of a clock. He who believes the clock is animate attributes the glory of the maker to the thing made. In this machine all of the movements flow

from one very simple magnetic force just as in a clock all the motions flow from a simple weight. (Olson 1990, 28)

In 1628, the English physician William Harvey envisioned the chambers of the heart moving "like the clacks of a water bellows." Nine years later, René Descartes described the way in which the parts of an animal's body respond to external stimuli in terms of the movement of the parts of simulacra (moving machines fabricated to imitate living beings). Indeed, he argued that those who have seen such machines "will look upon the body as a machine made by the hands of God" (Descartes 1955, 1: 116).

Initially, such explanations were self-consciously analogical. Living beings and the universe were understood to be *like* automata and clocks, but they were not identified with them. But this soon changed. By the end of the seventeenth century, a large fraction of all natural scientists in Europe probably agreed with Gottfried W. Leibniz when he wrote: "A body is never moved naturally except by another body which touches it and pushes it. Any other kind of operation on bodies is either miraculous or imaginary" (Olson 1990, 22). According to this point of view, all natural phenomena must be explained in terms of mechanical impacts, or they are not legitimately explained at all.

There were many varieties of mechanical philosophy in the seventeenth century. Some, including the varieties offered by Descartes, Thomas Hobbes, and Leibniz, insisted that some features of nature could be known by reason prior to any experience. Their authors generally argued that only knowledge derived from unimpeachable first principles and not from experience could legitimately be called scientific. Other mechanical philosophies, such as those developed by Robert Boyle and Pierre Gassendi, were grounded in experience, which could give only probable knowledge. Their authors generally denied that any knowledge of nature could achieve the kind of certainty sought by Descartes and Leibniz. Some mechanical philosophers argued that matter was defined by extension, from which it followed both that there could be no void space in the world and that such unextended entities as soul or spirit were radically separate from matter. Others saw matter as composed of small impenetrable chunks moving about in a great void and tended to think of spirit and soul as composed of extremely small, or "subtle," bits of matter. Regardless of where they stood regarding the fundamental source of scientific knowledge or the precise character of "matter," all mechanical philosophers insisted that only the contact of bits of matter with one another could produce any kind of natural change.

Catholic and Protestant scholars alike adopted almost every version of the mechanical philosophy, but whereas Protestant scientists seemed to dominate late seventeenth century developments, Catholics outside of the Dominican and Jesuit orders played a more important role in the first half of the century. Furthermore, for these Catholic mechanical philosophers, religious issues played a significant role in their theory choices and in their careers, usually in positive—but sometimes in negative—ways.

The first major Catholic thinker to promote a far ranging mechanical philosophy was the Minim friar Marin Mersenne. In the 1620s, Mersenne became disturbed by the threat of atheism, which he saw as coming primarily from Renaissance neo-Platonist scientific thought. In particular, he argued that by filling the world with hidden sympathies and occult forces, Hermetics, alchemists, and natural magicians blurred the distinction between the natural and the supernatural and undermined belief in the miracles that seemed so important to him as a warrant for Christian belief. Unable to accept the Thomist doctrines of the Jesuits and Dominicans because of a thoroughgoing commitment to nominalism or "mitigated skepticism" in philosophy, Mersenne denied the possibility of attaining absolutely certain knowledge of either natural or theological truth. Thus, he turned to empiricist studies in the sciences, which were at least able to provide reliable probable knowledge, and to the plausible emerging mechanistic accounts of phenomena that had appeared in the works of a number of medical writers and in *The Assayer* by Galileo.

In 1634, in a three-volume musical treatise entitled *Harmonie Universelle*, Mersenne first articulated a notion that was to become important for Descartes, Gassendi, and subsequent Catholic responses to virtually every form of scientific naturalism—including evolution—into the twenty-first century. Mersenne analyzed the responses that animals have to sound, as when a female bird responds to the male's song. He argued that animal responses to sound are best understood in terms of mechanical impressions on their senses: "[O]ne can say that they do not act so much as they are acted upon, and that objects make an impression on their senses such that what follows is necessary, just as it is necessary that the wheels of a clock follow the weight or spring that drives them" (Olson 1990, 35).

He then went on to claim that humans respond in a different way because God has imbued humans with a special kind of spirit or mind capable of considering the nature of sounds and of choosing their response to them. Whether Mersenne first published an idea that Descartes had suggested to him, or whether it was Mersenne from whom Descartes derived the notion, is uncertain. Both men were responding to the fundamental humanist separation of man, created in the likeness and image of God, from the rest of the natural order. As developed further by Descartes and Gassendi, naturalistic approaches were extended to all physical phenomena associated with humans and were abandoned only in connection with the rational soul or mind, which God added directly to the human body. According to Gassendi:

> The human soul is composed of two parts: . . . the irrational, embracing the vegetative and sensitive is corporeal, originates from the parents, and is like a medium or fastening (nexus) joining reason to the body; and . . . reason, or the mind, which is incorporeal, was created by God, and is infused and unified as the true form of the body. (Osler 1994, 66)

This is precisely the position taken by the Catholic Church today with respect to evolution by natural selection: it is deemed completely appropriate to the understanding of all living organisms other than man and to physiological changes in humankind, but not to the human soul, which is understood to be immediately infused by God into each person.

René Descartes did not initiate his philosophical discussions primarily for religious reasons. Instead, he was led to religious questions both because they helped to resolve critical problems that arose in his philosophical system and because he realized that unless he could account for certain religious doctrines in terms of his philosophy, that philosophy would never gain a significant hearing within Catholic Europe. Cartesian philosophy, as Descartes' beliefs came to be known, seems to have had at least three major sources. One lay in the mathematical education Descartes received from his Jesuit teachers. Mathematics offered Descartes a unique model of certain knowledge, both in terms of its deductive structure and in terms of the clarity and distinctness of its fundamental assumptions. The second lay in a set of life-transforming experiences Descartes had while visiting the gardens at Saint-Germain-en-Laye in 1614. There he came into contact with several spectacular automata created by the Francini brothers and began to think that perhaps living beings were nothing more than very complex automata. Finally, the revival of ancient skepticism, which was one of the products of the humanist search for ancient texts, had a major impact on Descartes' thought.

Descartes proposed to begin his philosophy by doubting everything that he knew, following the ancient skeptics, but then he argued that even while trying to doubt, he realized that for the very process of doubting to be possible, he had to exist as a thinking being. His famous line was "Cogito ergo Sum" usually translated as "I think, therefore I am." But how could one move beyond this single truth? The statement seemed to be true because it was clear and distinct. But what can guarantee that what we perceive as clearly and distinctly true is not just a deception? Descartes argued that only the existence of a God who is truly good and therefore not a deceiver can offer such a guarantee. Given the existence of God to be the guarantor of clear and distinct ideas, Descartes went on to develop his mechanistic physics in the *Principles of Philosophy* of 1644. It is clear, then, that a theological claim regarding the existence and character of God lay at the heart of Cartesian philosophy.

The radical separation of Cartesian matter, which was defined as pure extension, from mind or soul, which was defined as pure thought, raised a number of philosophical and theological questions. One crucial question was how, if in any way, the soul, which was traditionally thought of as an internal principle of self-movement of a substance, could interact with the body. Another was how the mind or soul related to God, since its cognitive ability seemed independent of all else for Descartes.

Descartes quite clearly understood that his views would seem threatening to traditional Catholic authorities; so he was very careful to be inoffensive. After the 1633 trial of Galileo, he withdrew two works that had been readied for publication, *The World* and *On Man*, to avoid Church disapproval, allowing their publication only after his death. He dedicated his *Meditations* to the dean and doctors of the Sacred Faculty of Theology at Paris. He tried to respond to concerns about the doctrine of the Eucharist by offering his own account of how the qualities associated with bread and wine could continue after they had been changed into the body and blood of Christ. (In his view, as long as the external surfaces remained the same, one would continue to see and taste bread and wine even though they were really the body and blood. Unfortunately for Descartes, this explanation seemed unsatisfying to virtually everyone.) Finally, in a textbook edition of *Principles of Philosophy*, Descartes included as his final principle, "I submit all my views to the authority of the Church" (Lennon 2000, 147). None of these attempts, however, quieted the criticisms, and in 1663, when *On Man* was finally published, Descartes' entire body of work was placed on the Index of Prohibited Books at the urging of the relatively liberal Jesuit Honore Fabri. Alas, Descartes' philosophy was formulated in such a way that it could not be viewed as merely hypothetical.

Descartes' philosophy did continue to find support among Catholics in spite of the Index and local condemnations by theology faculties at Louvain, Paris, Angers, and Caen. But there is evidence in individual cases that Church opposition to Cartesianism did have an impact. Jacques Rohault, for example, who had been among the foremost Cartesians in the early 1660s and the author of the most widely used Cartesian textbook, withdrew his support after a 1671 warning by the archbishop of Paris (Ashworth 1986, 152).

The third major Catholic mechanical philosopher was the secular priest Pierre Gassendi. Like Mersenne, Gassendi came to a mechanical philosophy in the 1620s as a result of his simultaneous rejection of neo-Platonic magical philosophies and Thomist-Aristotelianism based on revived skeptical arguments. More of a humanist scholar than either Mersenne or Descartes, Gassendi sought for an alternative ancient philosophy that would be more satisfactory than the others. He found this philosophy in ancient Epicurean atomism, which had been recovered through Lucretius' *De Rerum Naturae* and Diogenes Laertius' *Lives of the Philosophers* in the sixteenth century. The latter offered direct excerpts from Epicurus' writings while the former offered an extended poetic version of Epicurean natural science.

Though Epicurean notions had to be slightly modified to eradicate their original atheistic tendencies and suit them for Christian life, one feature made ancient atomism vastly more acceptable to Catholic scholars than, for example, Cartesianism. Epicurus had insisted that knowledge claims be grounded in empirical evidence and that they could only be probable, never certain. This position made it possible for Jesuits to embrace atomist-based mechanical philosophies as plausible hypotheses without insisting on their truth.

In Epicurus' version of atomism, the universe originated at some indefinite time in the past as an infinite number of small, hard, indestructible particles of various shapes moving "downward" through void space in parallel. Both to get the process of interactions among bodies that we see in the world started and to avoid the specter of complete physical determinism, Epicurus claimed that every once in a while at random intervals, a few particles spontaneously undergo a "clinamen," or gentle swerve in this downward path, initiating a collision with some other particle moving downward. In this way, a cascade of collisions is initiated which leads to the world as we know it. Ours is a world in which bodies composed of many atoms entangled with one another move in a variety of directions and interact with one another through collision (though in general, they continue to fall "down").

Gassendi modified Epicureanism by insisting that God created only a large but finite number of atoms initially and that he gave them their initial motions, which eventuated in their interacting according to his divine plan. As a consequence, Gassendi revived an early Christian tradition that went back to Clement of Alexandria and the Cappadocian father Basil the Great—that is, the so-called argument from design, or natural theology. If God created and guided the natural order by his "ordinary providence," then one should be able to detect characteristics of the divine by studying the natural order, which reflects his providential plan. Some Catholic authors took up natural theology, arguing, for example, that one could deduce God's wisdom and power even from the structure of insects. But it was in the hands of several English scholars who were at least indirectly influenced by Gassendi's works that the design argument and natural theology became a major constituent of modern Protestant theology.

Gassendi rejected the spontaneous gentle swerve of Epicurus and replaced it with the notion that an all-powerful God could occasionally—but rarely—interrupt the natural order, which expressed his ordinary providence, by exercising his "special providence" to create a miracle. Finally, as we have seen, Gassendi insisted that while the rest of the universe might be understood entirely in terms of matter in motion, the human soul was a direct, immaterial addition to each person from God. In order to preserve the Epicurean emphasis on pain avoidance and pleasure seeking, however, he argued that God imbued each soul with a desire for pleasure at the same time that he granted it freedom to pursue pleasure as it sees fit. The entire system of Christianized atomism appeared in its fullest form in Gassendi's posthumously published *Syntagma Philosophicum* in 1658, and it spread throughout Europe with relatively little serious opposition in spite of the well-founded fear of some that atomism might encourage others to think that once the universe had been created, it continued to move without any further divine intervention. It was picked up even before Gassendi's finished version appeared by the English natural philosopher Walter Charleton, through whose *Physiologia Epicuro-Gassendo-Charltoniana: or a A Fabric of Science Natural, Upon the Hypothesis of Atoms* of 1654 it entered the thought and writing of Robert Boyle, John Locke, and Isaac Newton, among others. As we shall see in the next chapter, Charleton embraced Gassendi's views in large part for the same reason that they could be utilized by the Jesuits—because the claims of atomist-based mechanical philosophy were appropriately modest in the face of God's infinite intellect and man's very finite one.

THE SPECIAL CASE OF BLAISE PASCAL

Perhaps the most famous French Catholic scientist and religious figure of the seventeenth century was a lay Catholic who rebelled against the authority of the Jesuits, accepted a version of the mechanical philosophy, and joined the Jansenist movement—a movement among French Catholics that emphasized the importance of God's grace over that of good works. Blaise Pascal was born in 1623, and was home-educated by his lawyer father. During the first part of his career, he worked in mathematics and designed the first working calculating machine in 1642. In addition, he did a series of barometric experiments which supported Evangelista Torricelli's anti-Aristotelian claim that the void exists. In 1646, he became a convert to the religious views of Cornelius Jansen, and then in 1654, he had a mystical experience of God that focused his later career on defending the Jansenist version of Christianity, though he did continue doing some mathematical work until shortly before his death in 1662.

Using his mathematical knowledge of the relationships between finite and infinite numbers, Pascal argued that while reason might be able to establish that God exists, it could not establish any of the characteristics of an infinite divine being, just as it could not establish such characteristics as the oddness or evenness of an infinite number. Only a direct, intuitive acceptance of God could guarantee belief, and such an acceptance was an issue for the heart, rather than for the head. Nonetheless, Pascal did offer what he considered to be a rational argument for opening one's self up to a belief in the Christian notion of a God who offered salvation through Christ—an argument often called Pascal's wager.

If such a God does not exist, and one believes, one might gain nothing, but neither does one lose anything. If, on the other hand, such a God does exist, belief stands to gain an infinite life of infinite happiness, while non-belief might lead to an infinite extent of infinite pain. In this case, if there is even the smallest chance that God exists, one should bet that he does because there is no downside to such a bet, while there is a huge potential upside. This argument, of course, assumes that there are only two options—the Christian God, or no God at all; but given this assumption, Pascal's wager can be understood as initiating a branch of mathematics known as game theory.

When we explore science as it developed in Catholic Europe from the mid-sixteenth century to the mid-eighteenth century, we see a very different picture than that of the intellectual scorched-earth portrayed by William Draper, A. D. White, and their followers. Certainly there were reactionary Catholics—especially among Dominicans—who resisted any challenge to Aristotle as interpreted by Saint Thomas. But the Jesuit order, in spite of an unwillingness to commit to the absolute truth of scientific theories, was tremendously active in promoting astronomical knowledge, experimental natural philosophy, and science education—especially in the mathematical sciences. Moreover, lay Catholics such as René Descartes; secular priests such as Pierre Gassendi; and Catholics within several smaller orders, including the Minim friar, Marin Mersenne played the leading role in promoting the rise of the mechanical philosophy in early seventeenth-century Europe. Mersenne also established an extensive network of scientific correspondents among Catholics and Protestants alike, which provided both support and challenges to stimulate and encourage the development of science and scientific institutions.

Sometimes, as in the case of Jesuits, Catholics' engagement with scientific activities was directly related to the institutional roles played by Catholic scholars within the church. At other times, as in the case of Descartes, religious implications were apparently drawn from science primarily as a way of legitimizing the science. But often, as in the cases of Mersenne, Gassendi, and the dissident Catholic mathematician and physicist Blaise Pascal, there were important religious issues that drove Catholic scholars to develop their particular scientific perspectives. Both Mersenne and Gassendi adopted the mechanical philosophy because it implied the need for a transcendent divine author of the universe in a way that neo-Platonic and Hermetic natural philosophies did not, and Pascal developed the beginnings of game theory largely to find intellectual support for his emotional conviction that God exists.

Chapter 4

Science and Religion in England, 1590–1740

There is general agreement that the rise of Britain as a center of scientific activity during the seventeenth century was related both to the growth of Britain as a commercial power and to the positive environment that British religious developments created for the pursuit of natural knowledge. But there has been vigorous debate about which religious groups were most hospitable to scientific activity. In this chapter, I will argue that British enthusiasm for experimental natural science was strongly encouraged among both Puritans and moderate Anglicans during the seventeenth century, but that each religious community tended to favor a particular form of science for its own special reasons. As a consequence, religious considerations played a significant role not just in the choice to become engaged in scientific activity, but also in the choice of what practices to engage in, what theories to support, and what topics to pursue.

With a few exceptions, Puritans and other radical religious sectarians promoted alchemy both for its presumed medical and commercial utility and because Paracelsan alchemy, in particular, fit well within their millenarian view of the world. Moderate Anglicans, on the other hand, tended to favor variants of the mechanical, or corpuscular, philosophy, in part because its admittedly hypothetical character promoted a kind of intellectual modesty and tolerance that they hoped would transfer to the religious realm. In addition, they were often concerned that the naturally animated universe associated with alchemy and the other occult sciences obviated the need for an immanent God and encouraged pantheistic tendencies (i.e., the worship of the world rather than its transcendent creator). Though not disinterested in the application of scientific knowledge to the improvement of the human condition, Latitudinarians, or liberal Anglicans, were more often drawn to search for natural knowledge for reasons associated with the tradition of natural theology because they expected that contemplation of the natural world, or God's "second book," could lead them to an understanding of God's characteristics and of their own Christian duties.

For the most part, after the restoration of the monarchy in 1660, High Church, or conservative Anglicans accepted neither the millenarian tendencies and the emphasis on inner light associated with the more radical sectarians, nor the emphasis on probabilism and natural theology associated with the Latitudinarians. As a consequence, though a few High Church figures could be found among seventeenth-century alchemists and mechanical philosophers, there was a growing distrust of the sciences in general among conservative Anglicans that manifested itself in the late seventeenth and early eighteenth centuries as a generalized hostility to the natural sciences or as interest in alternative, Scripture-based natural philosophies such as that articulated by John Hutchinson in *Moses' Principia* of 1737.

THE ANGLICAN FOCUS ON NATURAL THEOLOGY

Paul's first letter to the Romans states that "All that may be known of God by men lies plain before their eyes; indeed God has disclosed it to them. His invisible attributes, that is to say, his everlasting power and deity, have been visible, ever since the world began, to the eye of reason, in the things that he has made" (Romans 1:19–20). This notion that the natural world provides clues to God's nature—that there might be a natural theology—was repeated by Clement of Alexandria, Origen, and by Saint Basil the Great (329–379) in whose hands it became the foundation for a genre of Christian writing—the "hexameron," or commentary on the six days of creation—that persisted through late antiquity, the Middle Ages, and well into the early modern period. Even the Protestant reformers Martin Luther and John Calvin wrote commentaries on Genesis.

For most of the Greek-speaking fathers who initiated the tradition of natural theology, study of the natural sciences was not necessary for salvation. It was an activity that was encouraged as a supplement to faith in Scripture intended principally for what we would call the intellectual elite and what Clement called the "Christian Gnostic." The class of learned Christians could somehow enrich their religious experiences through the study of nature. This position was reiterated by John Calvin in *Institutes of Christian Religion* in 1536, and it continued to encourage natural theological writings on the Continent through the eighteenth century. But natural theology took on a much more important role for Anglicans around the beginning of the seventeenth century.

If we turn to England in the seventeenth century, we see a striking rise in the number of works devoted to natural theology and in the readership for those works. One possible explanation for this rise is that English scholars, independent of any religious reasons, became interested in the natural sciences. Then, in order to justify their interests within a society that still took religion seriously, they turned to natural theology. This account fails, however, to explain why similar explosions of natural theologizing did not occur elsewhere as interests in natural science grew, nor does it account for the timing of the growing English concern with natural theology, which tended to lead the growing interest in the sciences rather than to lag behind it. Instead, a special interest in natural theology developed in England initially within the Anglican faith as a response to a unique set of circumstances facing the Anglican Church under Queen Elizabeth I, and that interest in turn stimulated interest in certain forms of contemporary science.

When Henry VIII (1491–1547) established the Anglican Church in 1537, the rejection of Catholicism involved more than a transfer of Church property to the Crown and the removal of barriers to Henry's divorce from Katherine of Aragon. English Christian humanists, following the lead of John Colet and Erasmus, took hints from the Protestant Reformation, including the renunciation of relic cults and indulgences. But the new Anglican Church retained an emphasis on the sacraments, on a rich musical and Latin liturgical worship, and on the special mediating role of the priesthood and the ecclesiastical hierarchy.

During the brief reign of Edward VI (1537–1553), the tendency toward reform proceeded slowly. The Bible was translated into English, but private reading was discouraged, and the English-language *Book of Common Prayer* became the foundation for services. Younger clergy pushed for more radical reforms, denouncing the ornate rituals and the ecclesiastical hierarchy of the church; but before they could come into the ascendant, Edward died and Mary, the Catholic daughter of Henry VIII and Katherine of Aragon, returned England to Catholicism. Some vocal Anglican leaders, including Archbishop Thomas Cranmer, were executed. Other leaders, including John Whitgift, vice chancellor of Cambridge University, stayed home and kept a low profile. But nearly 800 Anglican intellectuals went into exile in Switzerland and Germany. These "Marian exiles" tended to absorb Calvinist ideas abroad and to live for the day when Mary would die and they could come back to England and continue to reform the Anglican Church.

Elizabeth I (1533–1603) came to the throne in 1558, bringing most of the exiles home; but she was caught in a very difficult position because she could not rule effectively without the support of both the Catholic-leaning aristocracy and the merchant and middle classes who leaned toward radical Protestantism. Whether out of political shrewdness, an egocentric desire to keep control of the church in her own hands, or some other motive or combination of motives, Elizabeth managed to maneuver Parliament and Convocation (the assembly of Anglican clergy) into accepting a very moderately reformed church, the chief features of which were that she was acknowledged as supreme governor of the church and the hierarchy of priests, bishops, and archbishops was retained. On the other hand, bishops were not allowed to require any specific doctrinal commitments from the lower clergy beyond the acknowledgment of Elizabeth's supremacy; so there was room in the church for those with strong Calvinist leanings in theology. Indeed, it is likely that up to 80 percent of the lower clergy sought further reforms. At the higher level, Elizabeth continued her inclusive policy by appointing six bishops who had actually served in the Marian church, six who had stayed home but refused church office under Mary, and fourteen who had gone into exile. While her policies were designed to keep religious moderates in the highest church offices, they were not intended to drive either religious conservatives or religious radicals out of the nation or the church.

For the most part, Catholics and crypto-Catholics were relatively pleased with Elizabeth's policies. The more radical Calvinists, led by Thomas Cartwright and a young man named Walter Travers, however, pushed aggressively for greater reform. They fought to replace the ecclesiastical governance of the church with a Presbyterian form of governance; they pushed for abandoning the *Book of Common Prayer* and liturgical worship in favor of services centered on preaching; they insisted upon justification by faith alone without regard to good works; and they insisted upon the primacy of Scripture to such an extent that they argued that anything not explicitly commanded in the Bible was to be considered sinful. Those who sought these reforms came to be called Puritans, and in 1572, they focused their efforts in *An Admonition to the Parliament*, demanding that such "Popish abominations" as the mediating role of priests and the use of baptismal fonts and ornate vestments be abolished.

After a brief and unimpressive attempt to respond to Puritan demands, Elizabeth's Archbishop of Canterbury (the clerical head of the Anglican Church), John Whitgift, picked a young Oxford-trained scholar, Richard Hooker (1554–1600), to create a doctrinal defense of Elizabeth's moderate policies and of Anglican theology, liturgy, and governance. Moreover, this defense was to promote Elizabeth's policy of inclusion, to keep Puritans and conservative Anglicans alike within the church rather than to drive them out. To accomplish his monumental task, Hooker wrote *The Laws of Ecclesiastical Polity*. The first four books appeared in 1593, but although the rest were drafted very quickly after that, they were not released until later, because while Hooker was generally supportive of Elizabeth and Whitgift's goals, once he began his argument it led him in some directions not completely consistent with their wishes or those of Elizabeth's successor, James I. Thus the last four books appeared only as Anglican religious politics caught up with Hooker's ideas. One consequence of this staged appearance of Hooker's work was that it remained fresh and before the public for nearly two-thirds of a century.

For our present purposes, only the first book, which sets out Hooker's most general principles, is critical. In an attempt to avoid the extremes of Puritan scriptural exclusiveness on the one hand and Catholic dependence on the authority of church councils on the other, Hooker chose to emphasize the roles of human reason and natural theology in justifying religious beliefs and practices. De-emphasizing the Calvinist notion that salvation demands some special dispensation of grace, Hooker focused on elements of religion that were, in principle, accessible to all. Drawing heavily from Basil and Origen, he insisted that from an investigation of nature, "the minds of mere natural men have attained to know not only that there is a God, but also the power, force, wisdom, and other properties God hath, and how

All things depend on him" (1845, 1:176). Furthermore, he argued that we can learn our Christian duties from nature as well:

The knowledge of every the least thing in the world hath in it a second particular benefit unto us, inasmuch as it serveth to minister rules, cannons and laws, for men to direct those actions which we properly term human. (1845, 1:175)

Unlike the most extreme Calvinists on the Puritan left and Dominicans on the Catholic right, who insisted upon God's absolute moment-to-moment willfulness, Hooker believed that God constrained himself to act according to reason. Thus, he wrote, "they err, therefore, who think that of the will of God to do this or that, there is no reason" (1845, 1:151). The rationality of God's actions is what ultimately makes natural theology both possible and powerful. The fact that we can be assured that the laws that God imposed upon the natural order are subject to reason allows us to approach God through the investigation of his reason as manifested in the creation by using our own reason. This notion belied the Puritan's insistence that God's commands could only be known from Scripture, and Hooker took great pains to make this point very explicitly: "It is their error," he wrote, "to think that the only law which God hath appointed unto man . . . is the sacred Scripture" (1845, 1:224).

Indeed, Hooker went far beyond the arguments of the early church fathers, of Saint Thomas, or of Calvin by arguing that natural theology did not simply provide a desirable but unnecessary supplement to scriptural revelation for intellectuals. Instead, he argued that it was a necessary complement to Scripture for everyone:

There is in Scripture . . . no defect, but that any man, what place of calling soever he hold in the church of God, may have thereby the light of his natural understanding so perfected, that the one being relieved by the other, there can want no part of needful instruction unto any good work which God requireth, be it natural or supernatural, belonging simply to men as men, or to men as they are united into any kind of society. *It sufficeth therefore that Nature and Scripture do serve in such full sort that they both jointly and not severally either of them be so complete that unto everlasting felicity we need not the knowledge of anything more than these two may easily furnish.* (1845, 1:215–216; emphasis mine)

Both Luther and Calvin had admitted that humans must study natural law in order to get along in secular society in this world, but

Hooker was telling his readers that even full attention to Scripture was insufficient for salvation without the additional awareness brought to humans through natural reason and the observation of nature. Not only did this claim deny Puritan scriptural exclusivism, it also denied the Catholic insistence upon tradition. Between nature and Scripture, everything necessary was given to humankind. Tradition was thus redundant if it agreed with natural reason and Scripture, or it was erroneous if it did not.

Some Christian duties are certainly stipulated in Scripture. Others are to be discovered by natural reason applied to nature. Both of these kinds of duties are necessary to salvation and they constitute the minimum demands upon a Christian. But, according to Hooker, not all topics of interest to Christians are covered by Scripture and natural law. With regard to these neutral or "indifferent" topics, among which he includes the forms of church ritual, the use of music and vestments, and the method of church governance, rules may be established by the Church. All such "positive" rules or laws should be produced for good reasons in the context in which they appear, and all should be obeyed if the Christian community is to survive. It is at the level of the kinds of activities governed by these positive laws, Hooker argued, that the vast majority of divergences between Puritanism and moderate Anglicanism occurred; and throughout the bulk of books 2–8 of *The Laws of Ecclesiastical Polity*, Hooker attempted to show that, practice by practice and rule by rule, established Anglican practices were more reasonable than those proposed by the Puritans. At the same time, by insisting on the relatively non-essential character of those issues covered by positive rather than sacred or natural laws, Hooker sought to make it easy for Anglicanism to reincorporate those who erred with regard to positive laws. From Hooker's time to the present, Anglican natural theology has been associated with the position that a wide variety of religious practices can offer salvation even though for earthly reasons some forms of practice might be preferred.

On two additional issues, Hooker initiated lines of argument that were to have long-term impact on English religion and natural philosophy. The first of these was connected with the issue of what counts as an acceptable argument both in religious discussions and in natural science. Hooker diverged substantially from Thomist practice and introduced a doctrine that made Anglican natural theology particularly appealing to those who were voluntarists regarding God's will, as Hooker definitely was not. This doctrine goes under the name of

"probabilism." According to most Christian Aristotelian philosophers including Thomas Aquinas, one should demand absolute certainty of any religious or natural scientific statement that commands assent. Such certainty can be achieved in only two ways: through immediate intuition or through logical deduction from intuitively given first principles. In the *Laws of Ecclesiastical Polity*, Hooker followed other Reformation thinkers in shifting the focus of Christian doctrine from strictly theological issues—that is, issues concerned with the nature of God—to issues of Christian ethics, morality, and duties. Hooker recognized that once this step was taken, Christian doctrine took the form of an Aristotelian "practical" science, rather than that of a "theoretical" science. As a consequence, it became a subject for probable arguments rather than conclusive demonstrations.

In general, Puritans insisted that the truth of Scripture is guaranteed to each Christian believer by an intuitive "inner light" that is granted by God and is unchallengeable. Hooker, on the other hand, was concerned about the plight of "weak" Christians, presumably including himself, who did not feel intuitive certainty regarding either the authority or the meaning of Scripture. It is true, he admitted, that "the mind of man desireth to know the truth according to the most infallible certainty which the nature of things can yield" (Hooker 1845, 1:262). Thus, when intuitive or demonstrative proofs are to be had they must outweigh any number of probable arguments. However, "in defect of proof infallible, because the mind doth rather follow probable persuasions rather than approve the things that have in them no likelihood of truth at all," merely probable arguments must be accepted so long as they are "what a reasonable man would accept" (Hooker 1845, 1:263).

The best that one could hope for was what later Anglican thinkers, including William Chillingworth and John Locke, the greatest developers of Hooker's probabilist ideas, called "moral certainty," or enough confidence to preclude *reasonable* doubt. Most seventeenth-century followers of Hooker came to the conclusion that very little knowledge that purported to be about things that really exist—whether those things were natural objects of sensory experience, or non-sensed spiritual entities such as angels or even God—could be more than probable. One consequence of this probabilist doctrine emphasized by both Chillingworth and Locke was that the acceptance of religious and scientific propositions alike had to depend upon the same kind of evaluation of evidence and argumentation. A second consequence was that persons should be very cautious about imposing their religious opinions on others, because they could never be absolutely certain regarding the truth of their own views.

The second of Hooker's arguments that had a major impact on religion and natural philosophy had to do with God's ongoing role in the natural world. Hooker went well beyond Saint Thomas's concern regarding God as first cause of the world to emphasize God's roles as creator, organizer, lawgiver, and efficient cause of all that happens. "It cannot be," he wrote, "But nature has some director of infinite knowledge to guide her in all her ways. . . . Those things which nature is said to do, are by divine art performed using nature as an instrument; nor is there any such art or knowledge divine in nature herself working, but in the guide of natures work" (1845, 1:156–157). All natural objects (except voluntary agents) do "proceedeth originally from some such agent, as knoweth, appointeth, holdeth up, and even actually formeth the same" (Hooker 1845, 1:157). Hooker made two important and related points in these sentences. First, he denied the Hermetic/Paracelsan/neo-Platonist claim that the universe contains within itself an *anima mundi*, or world soul, capable of directing nature from the inside and without God. Second, he argued that God is more than the creator of the universe. He has an ongoing role to play, he is "both the Creator and the *Worker* of all in all" (1845, 1:158; emphasis mine).

Hooker was convinced that men do not know the precise way in which God acts in the world. Men simply label his unknowable disposition of events, God's "ordinary providence." Subsequent Anglican natural scientists, such as Robert Boyle, would see the investigation of God's ordinary providence as their central task.

THE PURITAN APPROACH TO NATURAL KNOWLEDGE

When the first four books of the *Laws of Ecclesiastical Polity* appeared in 1593, the most promising "new" scientific trends in England were associated with alchemy, astrology, and natural magic and with the names of Hermes Trismagistus, Paracelsus, Giordano Bruno, and John Dee. A small group around Thomas Harriot began to explore Epicurean atomism in the 1590s, and Francis Bacon briefly flirted with atomism between 1605 and 1612, but the mechanical philosophies that were to dominate mid- and late-seventeenth-century British science were yet to emerge as important.

Though there was a huge range of attitudes, practices, and theoretical perspectives expressed within the various Hermetic, magical, and occult philosophies of the late Renaissance, there was a set of common core assumptions, two of which were: (1) that the world is a living organism with an *anima mundi*, and (2) that it contains a variety of hidden forces which humans can learn to control as a gift from God and the planets. Clues regarding the use of these forces could be discovered in a variety of ways, from following microcosm-macrocosm relationships, from signs placed in objects by God, and by detailed experiments regarding the properties of different materials. Virtually all variants of occult and magical philosophy had deep and fundamental religious dimensions. In almost all of them, the acquisition of natural knowledge was seen primarily as a key to recovering the state of human material and spiritual perfection and the dominion over the world that had been lost with the Fall of Adam, and which was a prerequisite to the coming of the judgment day.

The most important variants of the occult and magical philosophies in Britain were associated with Paracelsan alchemy. This Paracelsan tradition did not reach its peak in Britain until the 1640s and 50s during the Civil War and Interregnum (see Figure 4.1); but Paracelsan ideas became important much earlier through the English translation of some of Paracelsus's prophetic and astrological tracts in 1575 as *Joyful News Out of Helvetia*. The Paracelsan emphasis on experimentation was furthermore promoted by numerous English authors, including John Dee and Thomas Moufet, who actually traveled to Basel to study Paracelsan medicine.

Although there were some features of Paracelsan natural philosophy and religion that might have appealed to moderate Anglicans, Paracelsan ideas were generally much more congenial to radical Protestant reformers. They were embraced by most Puritans with scientific interests, and they thus established positions against which later Anglican natural philosophy and natural theology were directed. In spite of its emphasis on experiment, Paracelsan natural philosophy was highly speculative and its author was nothing if not confident. Paracelsan philosophy thus seemed to many moderates to illustrate the worst characteristics of the "enthusiasm" that seemed to them to pervade the radical religious sects. According to the Anglican moderate Henry More, who was one of Isaac Newton's teachers at Cambridge, Paracelsus was a dangerous man, "whose unbridled imagination and bold and confident obtrusion of his uncouth and supine inventions upon the world has . . . given occasion to the

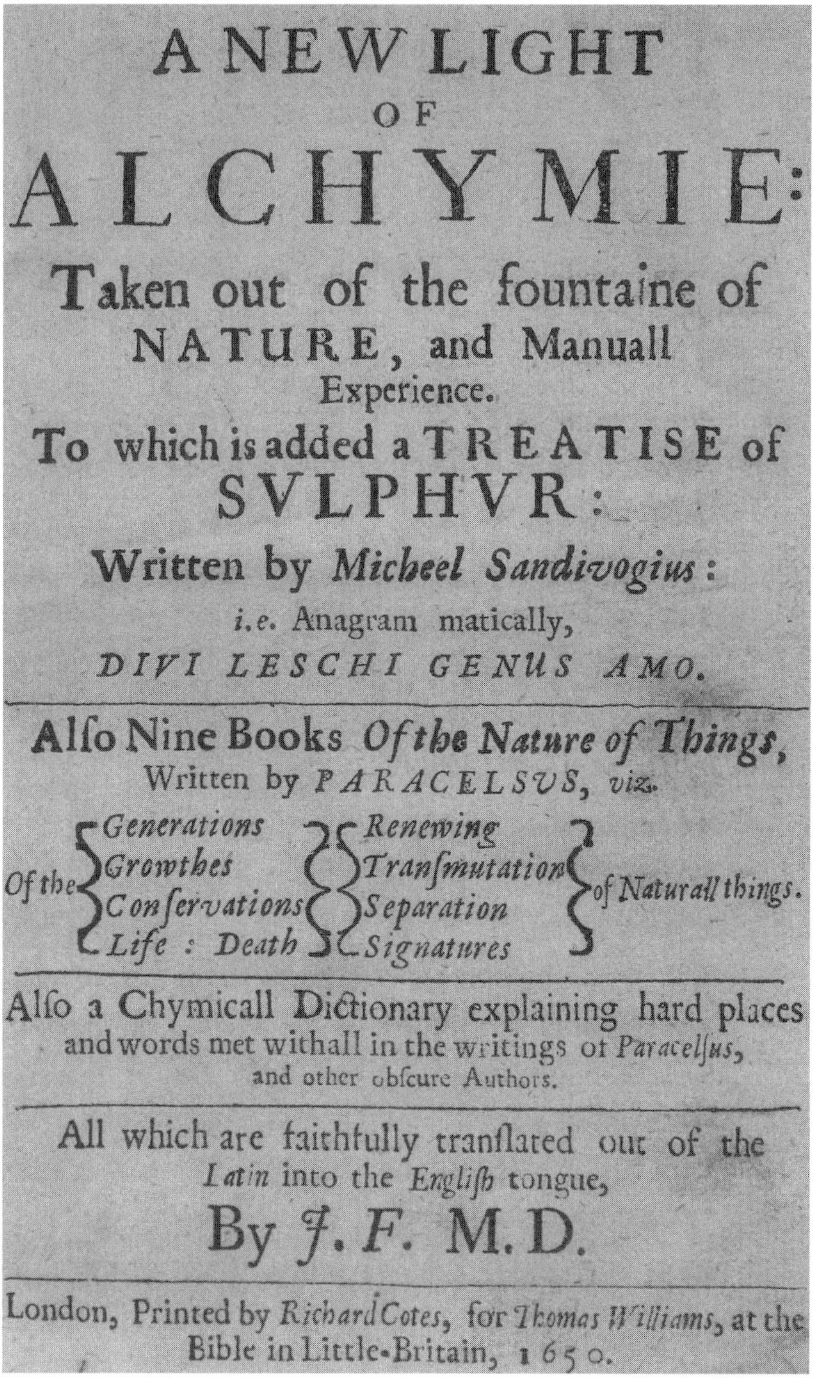

A NEW LIGHT
OF
ALCHYMIE:

Taken out of the fountaine of
NATURE, and Manuall
Experience.

To which is added a TREATISE of
SVLPHVR:

Written by *Micheel Sandivogius*:
i.e. Anagram matically,
DIVI LESCHI GENUS AMO.

Alſo Nine Books *Of the Nature of Things*,
Written by *PARACELSVS*, *viz.*

Of the { *Generations*, *Growthes*, *Conſervations*, *Life : Death* } { *Renewing*, *Tranſmutation*, *Separation*, *Signatures* } of *Naturall things*.

Alſo a Chymicall Dictionary explaining hard places
and words met withall in the writings of *Paracelſus*,
and other obſcure Authors.

All which are faithfully tranſlated out of the
Latin into the *Engliſh* tongue,

By *J. F.* M.D.

London, Printed by *Richard Cotes*, for *Thomas Williams*, at the
Bible in Little-Britain, 1650.

Figure 4.1. Title page from an alchemical compilation published by a Puritan printer during the period of Oliver Cromwell's leadership. Alchemical works multiplied during the Puritan ascendency. Courtesy of Special Collections, Honnold/Mudd Library.

wildest Philosophical Enthusiasms that were ever broached by any, either Christian or Heathen" (Webster 1982, 9).

Paracelsus and the moderate Anglicans also diverged over the adequacy of natural reason and the frequency of God's special intervention in the operations of the world. One of the main points of Hooker's emphasis on natural religion was that it did not depend on any special divine illumination or infusion of divine grace. Of course, human reason was a gift of grace; but natural reason was granted to all healthy humans, Christian or heathen, saved or not, as part of God's ordinary providence. Paracelsus, on the other hand, insisted that only through a special infusion of grace associated with salvation could true natural knowledge be achieved. Natural knowledge was thus the consequence of a direct illumination from God.

Perhaps the most powerful expression of the radical Protestant sense that any search for natural knowledge without God's saving grace through Christ was doomed came from Johann Andreae. In his *Civis Christianus* (*The City of Christ*) of 1619, Andreae described a vision of a Christian wandering aimlessly in the world. Totally exhausted, the wanderer turns to God in a desperate attempt to discover the meaning of his life:

In the half light there appeared a very small and humble sanctuary, having but a single portal, and to which only a darkling illumination was admitted through slits. Entering within, my sight at first failed me, but then came back little by little. Whereupon I saw various amateurish statues and paintings representing virtues, and they were half eaten away. . . . Then I saw, lying here and there, the demolished wheels, gears, and other parts of an armillary sphere—fragments of an ingenious piece of work which no master craftsman would be able to put back together. Also I perceived deplumed wings, broken ladders, and abandoned blocks and pulleys—vestiges of some great but vain project. *I found all in confusion and no one to tell me what it signified.*

O! The ineffable radiance that suddenly appeared! A light descended through the vault of this chapel—a light in human form, with flesh and body in every way like us, yet in splendor it was unquestionably God. His resplendence so illumined the place that what was heretofore hidden now stood revealed. The images of virtues, with marvelous artifice, reflected the light like purest crystal and even began to move as if they were alive; now they were whole and without any defect. *And the parts of the sphere came together to form a most exquisite timepiece; which portrayed with stupendous accuracy the movements of the universe and the plan of God's government. Everything displayed perfection.* (Montgomery 1973, 1:135–136; emphasis mine)

When Andreae's Christian looks within himself, he sees only the broken images of virtues and the flawed remains of human attempts to understand and modify the world. But when Christ enters his life, his virtues are perfected and he is able to recognize God's plan of the universe.

Just as the Paracelsans tended to focus on the special providence of God in bestowing knowledge on the reborn Christian natural philosopher, they tended to assume a wide scope for his special providential action in the natural world. God frequently intervened in the normal course of events, as when he created the star of Bethlehem. Furthermore, his interventions continue into modern times, both in the production of special signs—such as comets—in the heavens, and in the use of storms, wars, and epidemics to punish the wicked (Webster 1982, 21). Again, this aspect of Paracelsan philosophy fit well with the radical Protestant emphasis on the immediacy of God's retributive justice, but it violated the sense of stability and order that lay at the heart of moderate Anglican doctrine. Moderate Anglican natural philosophers and natural theologians did not deny God's ability to intervene miraculously in the world by suspending natural laws; but they did emphasize the extreme rarity of specially providential acts.

Yet another feature of Paracelsan doctrine that appealed to the radical reformers and set him apart from moderate Anglicans was Paracelsus' hatred of rituals and ceremonies—including the ringing of bells and burning of holy candles—which he interpreted as attempts at deceit or as forms of demonic magic (Webster 1982, 82).

Finally, and perhaps most importantly, Paracelsan doctrines reflected a social radicalism that grew in importance among religious reformers as the period of the Civil Wars approached. Opposed to corporate privilege and hierarchies of all kinds, Paracelsus had spent his entire life battling guilds of physicians and apothecaries. He persistently extolled the virtues of the poor, condemned the rich, and advocated a doctrine of service and poverty that approached pure Christian communism (Webster 1982, 52–54). Thus, during the 1640s and 1650s, as socially radical religious sects such as the Levellers and Diggers grew in importance among the revolutionaries, moderate Anglicans and even many moderate friends of the Presbyterians in Parliament felt increasingly threatened and anxious about the social implications of Paracelsan doctrines. In the next section, we will see that in several extremely important cases, moderate Anglican natural philosophers who had initially been drawn to Paracelsan alchemy converted to mechanical philosophies for a combination of social and religious reasons, and they sought explicitly to demonstrate that their new scientific perspectives offered support for the moderate and anti-enthusiastic version of religion advocated by Hooker and his supporters.

THE ORIGINS OF ANGLICAN MECHANICAL PHILOSOPHY

In 1652, the Anglican former alchemical advocate turned mechanical philosopher Walter Charleton (1620–1707) published *The Darkness of Atheism Dispelled by the Light of Nature: a Physico-Theological Treatise* in London. Two years later, he followed that work up with *Physiologia Epicuro-Gassendo-Charltonians, or a Fabric of Science Natural, Upon the Hypothesis of Atoms*. With these two works, Charleton ushered in a period during which many of the major English natural scientists, including John Wilkins, John Wallis, Robert Boyle, John Ray, and even Isaac Newton, wrote as much on religious topics as on scientific ones, and in which they all promoted the enterprise of natural theology. It was also a period when the mechanical philosophy, derived from ancient atomist roots and imported into England via Gassendi's works, came to dominate British natural philosophy. Finally, it was a period in which for most natural philosophers, religious attitudes and scientific works were intimately connected with one another.

Charleton was born in 1620 and studied medicine at Oxford. During the English Civil War, he was physician to Charles I, and Charles' death at the hand of the radical Protestant revolutionaries had a major impact on his life, turning him bitterly against the radicals, with whom he had previously had great sympathy. At the restoration, Charleton joined the Royal College of Physicians, which he served as president from 1689 to 1691; and when the Royal Society of London for the Advancement of Natural Knowledge was chartered in 1662, he became one of the founding members.

Until 1650, Charleton was an advocate of the Paracelsan "spiritualist" school of natural philosophy, which envisioned the world as animate and inhabited by a variety of immaterial occult forces. By 1652, however, he had become a confirmed mechanical, or corpuscular, philosopher. In his first major work, *The Darkness of Atheism*, he focused on how the mechanical philosophy could be used to defend Christianity against what he now viewed as the dangerous and atheistic spiritualists. According to the work's "Advertisement to the Reader," the overthrow of traditional civil and ecclesiastical authority had opened the way for the spread of "the most execrable *Heresies*, blasphemous *Enthusiasms*, Nay even professed *Atheism*" (Charleton 1652, n.p.), and his work was intended to stem this horrible tide.

For Charleton and for most mechanical philosophers, "God made, conserves, and regulates" nature (Charleton 1652, 154). As a consequence, the mechanical universe needed God both at the beginning and throughout time: "[T]his vast machine depends on God in every minute freshly to create it, or to conserve it in being by a continual communication" (Charleton 1652, 111). Furthermore, the world needs God to activate it. Just as a watch cannot run without a mainspring, so the world cannot run without God as an "energetical principle" or as the "spring in the engine of the world" (Charleton 1652, 216). In fact, from the perspective of many Anglican mechanical philosophers, only the mechanical philosophy could adequately support the theological claim that God's continuing providence is necessary to the ongoing operation of the world. Machines need to be driven by some external force; but living bodies do not. As a consequence, it was the Hermetic/Paracelsan view of the world with its presumption of an *anima mundi*, or source of motion integral to the world, that seemed to offer the specter of a world without God in the seventeenth century.

There was, however, a second problem with the animated universe of the Hermetics, Paracelsans, and other neo-Platonists. It seemed to invite worship of the creation, rather than the creator. Robert Boyle wrote in *On the Usefulness of Natural Philosophy* of 1662 that we should "not venerate the elements, the heaven, the sun, the moon, etc. These are but mirrors, wherein we may behold his excellent art who framed and adorned the world" (Boyle 1744, 1:441). Then he went on to explain why the mechanical philosophy that interprets the universe as an inanimate artifact encourages the most proper understanding of God's relationship to the universe:

When . . . I see a curious clock, how orderly every wheel and other part performs its own motions, and with what seeming unanimity they conspire to tell the hour, and to accomplish the designs of the artificer; I do *not* imagine that any of the wheels, etc., or the engine itself is endowed with reason, but commend that of the workman, who framed it so artfully. So when I contemplate the action of those several creatures, that make up the world, I do *not* conclude the inanimate species, at least, that it is made up of, or the vast engine itself, to act with reason or design, but admire and praise the most

> wise author, who by his admirable contrivance, can so readily produce effects, to which so great a number of successive and conspiring causes are required. (1744, 1:447)

More than twenty years later, in 1686, Boyle was still battling the notion of an animate universe filled with occult powers. In *A Free Inquiry into the Vulgarly Received Notion of Nature,* he returned in print to attack the sect of men "as well professing Christianity, as pretending to philosophy—who do very much symbolize [sympathize] with the ancient Heathens, and talk much indeed of God, but mean such a one, as is not really distinct from the animated and intelligent universe; but is, on that account, very differing from the true God, that we Christians believe and worship" (1744, 4:376).

Charleton and Boyle opposed those "spiritualists" who called on God's supernatural agency to account for unusual, but not extraordinary, phenomena, as when Johannes von Helmont described the rainbow as "a Supernatural Meteor, . . . having no dependence on natural causes" (Charleton 1652, 155). According to Charleton, God ordinarily operates in nature through secondary, natural causes which mediate between him and the world. Subsequent mechanical natural theologians spoke about God's "ordinary providence" in discussing what they viewed as God's acting through natural causes. But Charleton and others were also insistent that God could also act immediately and miraculously through his "special providence" without the interposition of natural laws. Charleton was particularly at pains to demonstrate that such miracles have actually taken place, in part in order to analogically justify the monarchical prerogative, which allowed the king to act outside the framework of civil law. Thus, he insisted that in spite of the fact that God established a "settled course" in nature, he

> hath not thereby so tied up his hands, or limited his Prerogative, as not to have reserved to himself an absolute superiority, or capacity, at pleasure to infringe, transcend, or pervert it, by giving special dispensation to any of his creatures, to vary the manner of their activities, in order to the causation of any effect, which his own prudence shall think expedient. (Charleton 1652, 129–130)

Sympathizers with Parliament (as opposed to Royalists like Charleton) emphasized the rarity of God's special intervention—even though they admitted the existence of miracles. Boyle, for example, agreed that God "is not overruled as men are fain to say viewing nature, by the head strong motions of matter, but sometimes purposely over-rules the regular [motions of matter], to execute his justice" (Jacob 1978, 228). But he was inclined to think that for virtually all purposes, the world was like a complicated clock,

> where all things are so skillfully contrived, that the engine being once set amoving, all things proceed, according to the artificer's first design, and the motions of the little statues that at such hours perform these or those things do not require . . . the peculiar interposing of the artificer or any intelligent agent employed by him, but perform their functions upon particular occasions, by virtue of the general and primitive contrivance of the Whole Engine. (Hall 1966, 152)

Moreover, he insisted that most of the seeming anomalies in the physical world that were seen by the spiritualists as special interventions of God were actually "the genuine consequences of the order . . . [God] was pleased to settle in the world; by whose course the grand agents of the universe were impowered and determined to act" (1966, 152).

On one final issue, Charleton's version of the mechanical philosophy illustrates a very critical aspect of it in connection with its relation to religion. In opposition to the radical sectarians who claimed the certainty of "divine illumination" for their science as well as their religion, Charleton insisted that the fundamental assumption of his physics—that is, that all bodies are composed of discrete small particles called atoms—is merely probable and not certain:

> [I]t is most *possible* and *verisimilous* that every physical continuum should consist of atoms; yet [it is] not absolutely *necessary*. For, insomuch as the true idea of nature is proper only to that *eternal intellect* which first conceived it: it cannot but be one of the highest degrees of madness for dull and unequal man to pretend to an exact, or an adequate comprehension thereof. We need not advertise, that the zenith to a sober physiologists [i.e., physicists'] ambition, is only to take the copy of nature from her shadow, and from the reflex of her sensible aspirations to describe her in such a symmetrical form, as may appear most plausibly satisfactory to the solution of all her phenomena. (1654, 128)

Moreover, Charleton argued, following Hooker's example, we face the same kind of uncertainty with regard to religion. Boyle also rejected the Paracelsan claim regarding divine inspiration in science; but he linked it to a critical religious argument. "I dare not affirm with some of the Helmontians and Paracelsans," he wrote, "that God discloses to men the great mystery of chemistry by good angels or nocturnal visions" (Jacob 1978, 16). When Boyle turned to religious issues, following Hooker, he insisted that religious knowledge, like natural knowledge, must be based on the evaluation of evidence, which can never produce absolute certainty. Boyle argued that this situation was important for the Christian, because certain knowledge forces assent. If there was definitive religious knowledge, we would be without choice in accepting Christian doctrine, and the notion of justification by faith would be completely meaningless. Given that religious knowledge is merely probable, however, its acceptance is a matter of choice (Boyle 1675, 97).

The probabilism associated with virtually all variants of experimental mechanical philosophy had a second consequence. It suggested that religious opinions, like natural philosophical ones, should be held less vehemently. Joseph Glanvill, one of Boyle's close supporters in the Royal Society made this point particularly clearly in *Scepsis Scientifica: or Confest Ignorance the Way to Science: In an Essay on the Vanity of Dogmatizing* (1662). "The Mechanik philosophy yields no security to irreligion," he insisted. Instead, it tends to "dispose mens spirits to more calmness and modesty, charity and prudence in their differences of religion" (Shapiro 1968, 34–35). Similarly, in the first official history of the Royal Society, published in 1702, Thomas Sprat wrote that the new experimental mechanical philosophy was producing "a race of young men . . . who were invincibly armed against the enchantments of enthusiasms" (1702, 53).

One of the crucial features of the mechanical philosophy as it had been explored in France by Descartes and Gassendi and in England by Charleton, Boyle, and most experimental natural philosophers, is captured in Boyle's *The Excellency and Grounds of the Mechanical Hypothesis* of 1674. In that work he insists: "The philosophy I plead for reaches but to things purely corporeal" (Hall 1970, 311). The acceptance of a radical distinction between matter and spirit, or between body and soul, was what allowed these mechanical philosophers to claim that their natural philosophy "yields no security to irreligion."

There was, however, at least one very important mid-seventeenth-century English mechanical philosopher who denied the distinction between matter and spirit and whose works stimulated a significant backlash against mechanical philosophies within the moderate Anglican tradition. This person was Thomas Hobbes (1588–1679). Known almost exclusively as a political thinker today, Hobbes was recognized as a serious natural philosopher during the seventeenth century, though his opposition to the experimental practices that dominated the membership of the Royal Society led to his exclusion from that body.

Hobbes' mechanical philosophy also emerged as a response to early seventeenth-century religious and political developments. With the accession of Charles I, and especially with his appointment of the aggressively orthodox William Laud as Archbishop of Canterbury in 1633, a conservative faction within the Anglican hierarchy attempted to purge Puritan sympathizers from their positions in the church and to insist upon conformity to a list of Thirty-nine Articles of the Anglican faith as a condition for holding even civil office. William Chillingworth, drawing heavily from Hooker's arguments, led a moderate opposition, but the Laudians gained a major voice in the government. Their attempts to repress non-conformity in turn stimulated increasingly strident opposition; so religious tensions and passions intensified during the 1640s, building to a crisis that was of major significance in bringing on the Civil War in 1642.

Hobbes was convinced both that divided authority of any kind was bound to lead to civil war and that religious pretensions of autonomous authority were, in the present context, the most dangerous of all. Thus, in 1641, he wrote to William Cavendish, Earl of Devonshire, in support of his attempts to diminish the power of the Anglican bishops. "I am of the opinion that ministers ought to minister rather than govern," he wrote. Then he continued, arguing that conflict "betwene the *Spiritual* and *Civill* power has of late more than any other thing in the world, been the cause of civill warres in all places of Christendom" (Shapin and Shaffer 1985, 311).

In his *Elements of Law* of 1631, *De Cive* of 1643, *Leviathan* of 1651, and *De Corpore* of 1656, Thomas Hobbes developed a completely materialist philosophy to attack all forms of religion that did not accept complete subordination to civil authority. For Hobbes, as for Epicurus, it was the false belief in spirits, made possible by ignorance of the causes of events, that gave the clergy its power; so the most effective way to fight the power of the clergy was to demonstrate that spirits, or "incorporeal substances," do not exist.

Finally, Hobbes adopted the Epicurean claim that if one could produce a natural philosophy that explained all phenomena—including those that the priests attributed to the agency of spirits—in terms of matter in motion, then one would have demonstrated that there is no need for the supposition of immaterial agents, spirits, or souls. Since Hobbes believed that his purely materialist mechanical philosophy could account for all phenomena, that it led to no false claims, and that immaterial causes were literally inconceivable, he was convinced that anyone who considered the matter rationally was bound to accept his natural philosophy and deny the existence of spirits.

Hobbes' challenge to virtually all traditional forms of Christianity was horrifying to Anglicans and to most non-conformists alike, for it undermined almost all of their fundamental beliefs, including the immateriality of God and the immortality of the human soul. He thus became one of the most hated and vilified of seventeenth-century intellectuals and the very symbol of the atheistic menace faced by Christians. But materialism also took on another troubling aspect when it was linked to democratic politics—much to Hobbes' dislike—by Richard Overton, one of the founders of the Levellers, a sect of radical Protestants that emphasized not only spiritual but also political and material equality. In *Man's Mortallitie* of 1643, Overton argued, like Hobbes, that everything is material and that the soul is only a term used to characterize qualities of matter that must perish when the body does. Moreover, Overton used this idea openly to attack organized religious institutions and to defend pure egalitarianism. In doing so, he drew additional criticisms to materialism.

In the process of making his argument that immaterial entities are not only unnecessary, but that they are also inconceivable, Hobbes had departed from ancient atomist traditions in adopting his definition of matter from René Descartes' mechanistic natural philosophy. By agreeing with Descartes that matter is defined by extention—that which is matter is that which is extended or which occupies space—Hobbes placed himself in conflict with most British experimental mechanical philosophers as well as with religious figures. Thus, opposition to Hobbes created an important common interest among the bulk of the English scientific community and religious apologists.

The ancient atomists had posited the existence of atoms and the void, claiming that atoms move freely through empty space (the void) until they collide with one another to produce the phenomena we experience. Descartes and Hobbes, on the other hand, insisted that it follows from their definition of matter that there can be no void, because every space has extension and therefore must contain matter. As Hobbes expressed this notion:

> The world . . . is corporeal, that is to say, body; and hath the dimensions of magnitude, namely length, breadth, and depth. Also every part of body is likewise body, and hath like dimensions; and consequently every part of the universe is body, and that which is not body is not part of the universe: and because the universe is all, that which is no part of it is nothing; and consequently no where. (Hobbes 1839, 672)

Precisely the same kind of argument that precluded immaterial spirits from the world precluded the possibility of a void.

The denial of the void was certainly less critical to British experimental mechanical philosophers in the mid-seventeenth century than the denial of spirits was to the clergy. Those who interpreted Boyle's air-pump experiments were inclined to think that once the air had been removed from the receiver of the air pump, what was left was a void, but given their probabilistic approach to knowledge, they were also willing to admit that it was just possible that when the air was removed from a container, some matter composed of extremely small particles remained. What appalled the experimentalists about Hobbes' denial of the possibility of the void was that he could not offer experimental support for his claims and, from their perspective, he held his position dogmatically and fanatically. While the conclusion of the Hobbesian theory might have been compatible with conclusions acceptable to the experimentalists, Hobbes' whole way of arguing challenged their conception of proper scientific method. The Hobbesian—pure materialist—version of mechanical philosophy was therefore as troubling to most English mechanical natural philosophers as it was to the Anglican clergy.

THE ANTI-MATERIALIST RESPONSE TO HOBBES

For some Christian thinkers, the new mechanical philosophies of the seventeenth century seemed to offer themselves as powerful antidotes to perceived anti-Christian tendencies towards the spiritualistic or pantheistic natural philosophies linked to Paracelsus. But for others, the atheistic menace represented by Hobbes seemed so great that they insisted there must be some active, spiritual agent implanted within nature to implement God's natural laws and to carry out his providential aims. Most of the mid- to late-seventeenth-century advocates of an active principle in nature saw themselves as operating within the Cartesian tradition of mechanical philosophy; for Descartes' mind-body dualism allowed in principle for the operation of spirit in the universe. But the anti-materialist Anglicans tended to diverge from Descartes on so many issues that their natural philosophy became one of a basically different kind.

In his long term impact on both scientific and religious developments, the second most important British seventeenth century anti-materialist scientist was almost certainly John Ray, whose work *The Wisdom of God Manifested in the Works of the Creation* (1691) (see Figure 4.2) has been characterized as "the book which more than any other determined the character of the interpretation of nature 'till Charles Darwin's time" (Raven 1951, 1:110). But Ray's natural philosophy and natural theology were heavily dependant on the writings of a group of authors who taught at Cambridge University—including Henry More, Benjamin Whichcote, and Ralph Cudworth—usually known as the Cambridge Platonists.

The Cambridge Platonists were even more liberal in their religion and placed even more emphasis on natural theology than Richard Hooker. Indeed it was the Cambridge Platonists who were first designated as Latitudinarians, or "persons of wide swallow," by their more conservative enemies. According to Cudworth, there are only three essentials in the true religion: first, one must believe "that there is a God, an omnipotent, understanding Being, presiding over all." Second, one must believe that this God is "essentially good and just," so there must be something prior to God that is "in its own nature immutably and eternally just and unjust; and not by arbitrary will, law, and command only." Finally, one must believe "that we are so far forth principles or masters of our own actions as to be accountable to justice for them, or to make us guilty or blameworthy for what we do amiss and to deserve punishment accordingly" (Cudworth 1845, 1:xxxiv). Most important for our purposes, Cudworth argued that all three of these principles can in principle be established by natural reason without appeal to Scripture (1845, 1:xxxv).

While the first of Cudworth's principles would probably have been accepted by virtually all Christians at the time, the second and third were both aimed against the Calvinists, who insisted that it was God's

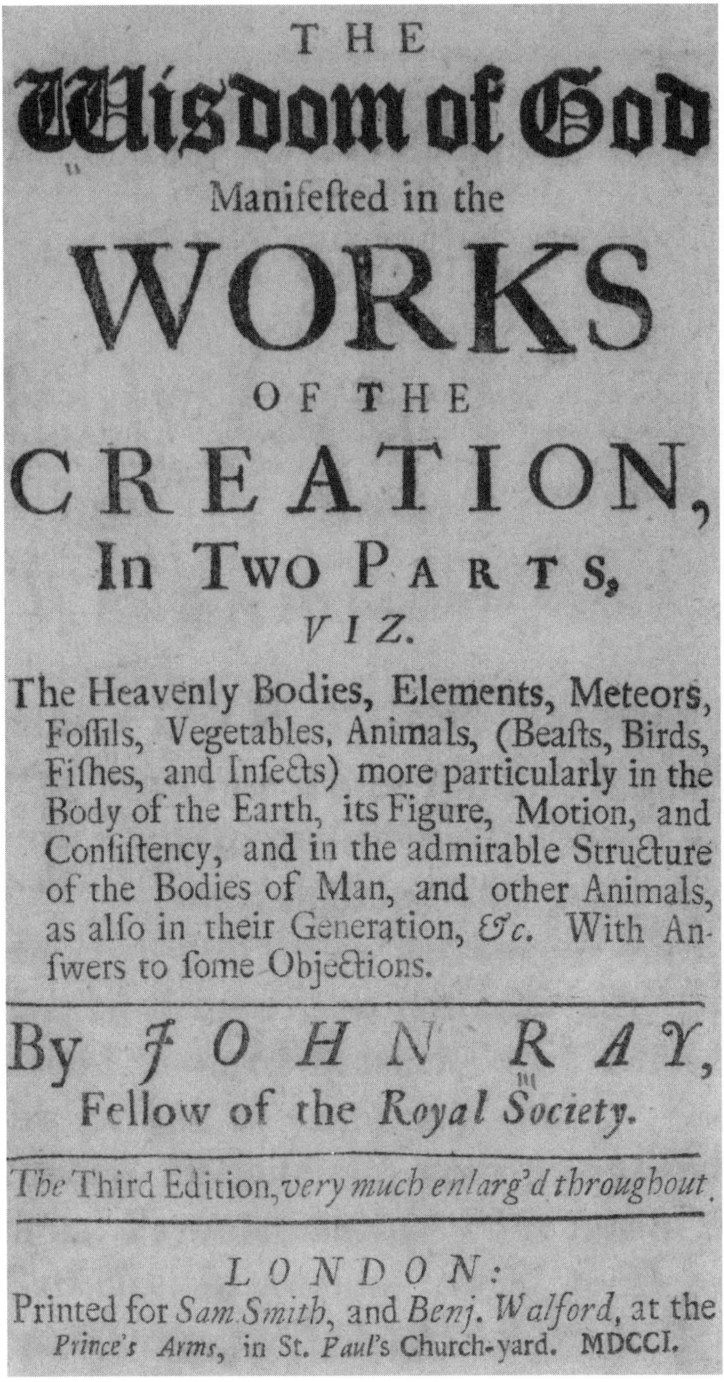

THE

Wiſdom of God

Manifeſted in the

WORKS

OF THE

CREATION,

In Two Parts,

VIZ.

The Heavenly Bodies, Elements, Meteors, Foſſils, Vegetables, Animals, (Beaſts, Birds, Fiſhes, and Inſects) more particularly in the Body of the Earth, its Figure, Motion, and Conſiſtency, and in the admirable Structure of the Bodies of Man, and other Animals, as alſo in their Generation, *&c.* With Anſwers to ſome Objections.

By *JOHN RAY*,

Fellow of the *Royal Society.*

The Third Edition, *very much enlarg'd throughout.*

LONDON:

Printed for *Sam. Smith*, and *Benj. Walford*, at the *Prince's Arms*, in St. *Paul's* Church-yard. MDCCI.

Figure 4.2. Title page from the third, expanded 1705 edition of John Ray's widely disseminated *Wisdom of God Manifested in the Works of the Creation.* Courtesy of the Claremont School of Theology.

will that made something good and that God's predestination of those who would be saved had nothing to do with their deservingness. It was equally aimed against the Hobbesians, whose pure materialist determinism seemed to deny the possibility of moral responsibility because it removed the possibility of human choice. On all three issues the Cambridge Platonists drew on a Platonist-Christian tradition that predated both the Reformation and the rise of Aristotelianism, and which went back to the writings of the earliest church fathers, including Origen and Clement of Alexandria.

In seeking to find a middle ground between the materialist atheists, "who derive all things from the fortuitous motions of senseless matter," and the bigoted Calvinists, "who needs [*sic*] have God to do all things himself immediately as if all in nature were a miracle" (Cudworth 1845, 1:606), Cudworth argued that there must be some nonmaterial but not intelligent "plastic nature" acting as a cause of at least some phenomena. Unless such a plastic nature exists, either "everything comes to pass fortuitously and without the direction of any mind or understanding, or else God himself doth all immediately, and, as it were with his own hands forms the body of every gnat and fly, insect and mite. . . ." (Cudworth 1845, 1:218) (see Figure 4.3). He dismisses the possibility of the random material cause of everything by appealing to the traditional design argument. It is, for example, not just probable but "so plain that nothing but sottish stupidity or Atheistic incredulity can make any doubt that eyes were made by Him for the end of seeing . . . and ears for the end of hearing" (Cudworth 1845, 2:616). He offers several inconclusive arguments for the existence of a spiritual agent mediating between God and the physical universe and then he presents two arguments, drawing parallels with Plato's *Timaeus*, which he considered to be conclusive, that God does not act directly to effect his ordinary providence. We will look at just the simplest of these arguments. Since God is perfect, he could never directly create a deformed organism. But many deformed organisms exist. As a consequence, God must govern the world through an inferior plastic nature which is "not altogether incapable (as well as human art) of being sometimes frustrated and disappointed" (Cudworth 1845, 1:223). Of course, as usual, God reserves to himself a special providence by which he might "supply the defects of [the plastic nature] and sometimes overrule it, forasmuch as this plastic nature cannot act electively, or with discretion" (Cudworth 1845, 1:224).

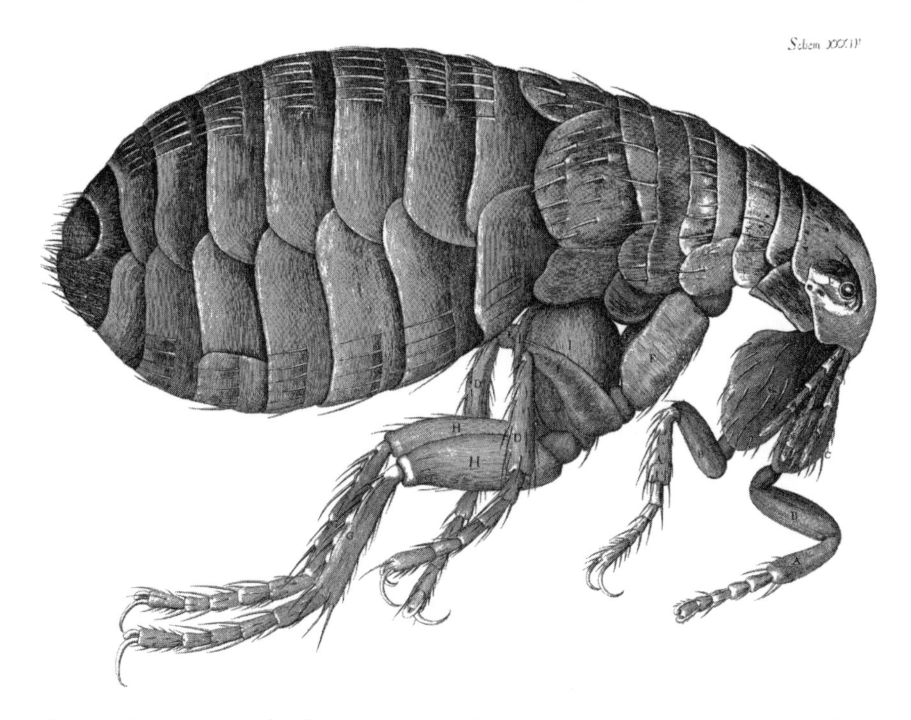

Figure 4.3. Drawing of a flea as observed through a microscope, fold-out plate from Robert Hooke's *Micrographia* (1665). Many natural theologians were impressed by the precision of God's handiwork in such creatures, in comparison to which John Ray wrote, "the sharpest needle doth appear as a blunt bar of iron" (Ray 1705, 66).

In the hands of John Ray (1627–1705), Cudworth's plastic nature became the foundation for a vitalist tradition in the discussion of living beings—later to be called biology—which was dominant throughout Europe during the eighteenth and early nineteenth centuries. Thus, it offers another important example of how religious considerations sometimes underlie scientific theory choice.

Ray received his divinity degree from Trinity College, Cambridge, in 1648 and remained there to teach until 1662 when he was expelled because he, like many church moderates, opposed and refused to subscribe to the Oath of Uniformity after the Restoration. Along with a friend, Francis Willoughby, Ray became interested in natural history and the problems of classifying different plants and animals. Willoughby died in 1670, but Ray continued their studies for over thirty-five years, publishing works on mammals, reptiles, birds, fishes, and insects, as well as three volumes on botany. In the process, Ray became one of the leading naturalists of his time, doing groundbreaking work in zoology, botany, and geology. Though he became an outstanding scientist, Ray never abandoned his initial commitment to divinity, and there is no doubt that throughout his life he considered his studies of nature to be a form of religious activity and as a way to glorify God. When asked to serve as secretary of the Royal Society of London in 1677, he declined on the grounds that divinity was his profession, even though his position on the Act of Uniformity made it impossible for him to serve officially in the Anglican Church. In 1691, he finally published his best-known work, *The Wisdom of God Manifested in the Works of the Creation*, in order to explain the religious implications of his scientific work. "Not being permitted to serve the church with my tongue in preaching, I know not but it may be my duty to serve it with my hand by writing," he began (Ray 1705, preface, unpaginated).

Like his Cambridge Platonist colleagues, Ray tells us that the first requirement of a Christian is to believe that God exists. Moreover he agrees that such a belief "*must be demonstrated by arguments drawn from the light of nature and works of creation*" (1705, 7: emphasis mine). Ray admits that there may be supernatural demonstrations of this first great truth grounded in inner light, in miracles, or in biblical prophecies. But supernatural demonstrations are "not common to all persons and times." They can thus be legitimately questioned by those who have not experienced them directly. The proofs that Ray will offer from natural phenomena are, on the other hand, "exposed to every man's view . . . [and are] intelligible to the meanest capacity" (1705, 7). Thus Ray continues Richard Hooker's special interest in the ordinary man who has never experienced an intense religious illumination.

He begins his central argument with a clear statement regarding the power of what came to be called "the argument from design":

There is no greater, at least no more palpable and convincing argument for the Existence of a Deity, than the admirable Art and Wisdom that discovers itself in the Make and Constitution, the Order and Disposition, the Ends and Uses of all the parts and members of this stately fabric of Heaven and Earth: for if, in the works of Art, as for example, a curious Ediface or Machine, Council, Design, and Direction to an End appearing in the whole frame, and in all the several pieces of it, do necessarily infer the being and operation of some intelligent Architect or Engineer, why shall not also in the Works of nature, that grandeur and magnificence, that excellent contrivance for Beauty, Order, Use, etc., which is observable in them wherein they do as much transcend the Effects of human Art as Infinite Power and Wisdom Exceeds finite, infer the Existence and Efficiency of an Omnipotent and All Wise Creator? (1705, 8)

While Ray might have admitted that the design argument does not offer absolute certainty, he did insist in his discussion of the design of the human eye for sight that it would be "absurd and unreasonable to affirm, either that it was not designed at all for this use or that it is impossible for man to know whether it was or not" (1705, 44).

Like Cudworth, Ray argued that there is an immaterial spiritual entity—a plastic nature—active in all living beings that accounts for their slow growth and for the occasional "bungles" and "errors" that were unimaginable if organisms were the direct and immediate creations of God. Ray, however, goes beyond these arguments to offer observational evidence for the action of non-mechanical, spiritual powers even in non-human nature. Starting with an extended discussion of the reasoning that his dog must go through in jumping from the floor to a chair, and then onto a table to reach an object that he wants, Ray goes on to discuss the way in which naturally timid fowl protect their young. Such phenomena, he claims, demonstrate the action of spirit in animals. Even more importantly, he argues that direct evidence for a vital principle or plastic nature in animals comes from the fact that the heart muscle will continue to contract even after it has been removed from the body of an animal.

At the hands of Ray's admirers, such as Albrect von Haller, who discovered that almost all muscle tissues demonstrate "irritability"—that is, the tendency to contract when touched—a "vitalist" physiology proved scientifically more fruitful throughout the eighteenth century than the mechanical tradition encouraged by Descartes and Boyle. No doubt there were many contributing reasons; but among the factors we must acknowledge that kept vitalist physiology alive was its conformity to a lively tradition of natural theology initiated by Cudworth and Ray.

There is no doubt that Ray, Cudworth, and their Cambridge Platonist colleagues were pious Christians who were convinced that the Bible constituted God's revealed word. But there is also little question that in their almost casual rejection of the adequacy of supernatural arguments for God's existence and wisdom, in their arguments that our belief in God must be grounded in "the light of nature and the works of creation," and in their emphasis on the design argument, these men moved away from more traditional forms of Christianity. By pushing to their extreme tendencies initiated in the writings of Hooker, the Cambridge Platonists seemed to encourage a natural religion that was increasingly divorced from Scripture and from any emphasis on the traditional mysteries of Christianity or the creeds and sacraments that gave religious practices their socially cohesive character.

Chapter 5

Newton's Religion, Newtonian Religions, and Eighteenth-Century Reactions

Among early modern scientists, Isaac Newton (1642–1727) was certainly one of the greatest and one whose work stimulated a great amount of religious commentary and controversy. But any reasonably honest and fair attempt to come to grips with Newton, Newtonian science, and related religious developments in the late seventeenth and early eighteenth centuries is immensely complicated for several reasons. It is complicated first because, as one of the foremost Newton scholars put it, "[Newton's] scientific discoveries and what Newtonians made of them, not his own religious utterances, helped to transform the religious outlook of the West . . . and in a way that would have mortified him" (Manuel 1974, 4). Second, Newton's own personal and private religious correspondence and unpublished or posthumously published religious writings have a very different thrust from his public religious pronouncements. Finally, the relationship of Newton's natural philosophy and religion to his serious alchemical concerns, to the mechanical philosophy of Boyle and Charleton, and to the anti-materialist philosophy of his Cambridge Platonist teachers was also extremely complex.

Newton's public religious statements almost uniformly reinforced the liberal Anglican trend toward ever greater emphasis on natural theology at the expense of scriptural religion. Yet privately, Newton's religion was primarily scriptural in its emphasis. Like the religion of the radical millenarian sectarians and that of modern fundamentalists, it was particularly concerned with the prophetic books of Daniel and the Revelation of John. Newton was convinced that the Jews rejected Christ because they had failed to understand their prophets, and he was convinced that his own age was one in which the prophecies were particularly important (Manuel 1974, 109). As a consequence, Newton devoted a tremendous amount of effort throughout his life to developing a system for interpreting prophecy.

Newton seems to have hidden most of his personal religious commitments to scriptural revelation from all but his closest friends, although his biblical scholarship led him to refuse to accept some of the Thirty-nine Articles of the Anglican Church and therefore to refuse to become an ordained Anglican priest. Since ordination was ordinarily a condition for becoming a fellow of one of the colleges at Cambridge (a requirement for being a member of the faculty), when Newton was offered the Lucasian Professorship in mathematics, his refusal to accept the Thirty-nine Articles necessitated a one-time-only waiver of the rules by royal dispensation. Until relatively recently, scholars have generally agreed that whatever Newton's private beliefs might have been, they had next to no historical impact either on his science or on subsequent religious developments. More recent studies, however, have suggested that his religious views did shape his science to some extent, and that his religious writings became important within religious discourse in two very different ways. On the one hand, they served as a direct and central source of modern Anglo-American fundamentalist theology. On the other, they served as a significant stimulus to the development of the higher biblical criticism.

Two themes did run through both Newton's public and private religious reflections. One was a passionate—some might say obsessive—concern with our duty to examine the evidentiary basis for both natural and scriptural religion. This obsession with evidence was tightly bound to his approach to natural philosophy. "The world loves to be deceived," he wrote. "There are but few who seek to understand the religion they profess—to examine whether it be true with a resolution to choose and profess that religion which in their judgement appears the truest" (Manuel 1974, 112). Newton proposed to show seekers how to undertake such an examination of both the natural and scriptural foundations for Christianity.

The second central theme of Newton's natural and scriptural religion alike was an emphasis on God's immanence, dominion, and absolute freedom of will. Newton's God was, like Hooker's, one who acted immediately and continuously in the world, both to sustain its orderly structures and to serve his own special ends. Newton was thus appalled when Gottfried Leibniz insisted that God must have "created the world so perfect that it can never fall into disorder or need to be amended," because such a world would not have an ongoing need for God's providential presence (Hall and Hall 1962, 584). And he was delighted when his most famous scientific work, *The Mathematical Principles of Natural Philosophy* (hereafter the *Principia*), seemed to demonstrate the absolute need for an occasional, special divine intervention to keep the solar system from collapsing. Newton acknowledged that the traditional argument from design provided a step on the path to knowing God as a creative intelligence, and he promoted the use of the design argument, offering new examples. But, he insisted in a famous "General Scholium" at the end of the *Principia*, "we *reverence* and *adore* him on account of his dominion; for we adore him as his servants, and a God without dominion, providence, and final causes, is nothing but Fate and Nature" (Newton 1962, 584; emphasis mine). So, if Newtonian arguments eventually led to an avid emphasis on God as a supremely intelligent creator without much of a continuing role to play in the world, that was certainly not what Newton had intended.

NEWTON'S SCIENCE AND REPUTATION

It was because of his spectacular accomplishments as a scientist that Newton was so admired and that his ideas on virtually any topic were eagerly sought. Moreover, that success was linked in the popular mind to a special emphasis on scientific method. Newton seemed to many to offer a new and powerful synthesis of older, rationalist traditions in science—which were modeled on Euclidean geometry and offered absolute certainty—with newer, experimental traditions, linked to alchemical practices, the writings of Bacon, and the works of Boyle and his friends in the Royal Society.

Born in 1642 into a prosperous Lincolnshire farm family, Newton attended a local grammar school in Grantham. After a singularly unsuccessful attempt at becoming a farmer, Newton went off to Trinity College, Cambridge, in 1661. Though Cambridge records for this period were burned in a fire, it is clear that Newton made several important contacts during his first years at Cambridge. He became a student and friend of Henry More and Ralph Cudworth, whose interests in Cartesian science, natural theology, and millenarian religious ideas he soon came to share. He also became a student and protégé of Isaac Barrow, the Lucasian Professor of Mathematics, and one of the most famous Latitudinarian preachers of his day. It was with Barrow that the young Newton began his mathematical studies and his work in experimental optics in earnest. When Barrow left Cambridge to further his clerical career in London, he effectively bequeathed the Lucasian Chair to his brilliant student.

In early 1665, the plague struck Cambridge, and Newton went home to the farm where he began a period of about eighteen months of intense work in mathematics and natural philosophy. Reflecting later on this burst of creative activity, Newton wrote:

> In the beginning of the year 1665 I found the method of approximating series and the rule for reducing any dignity of any binomial into such a series [he discovered what we call the binomial theorem]. The same year I found the method of tangents of Gregory and Slusius, and in November had the direct method of fluxions [he had invented what we call differential calculus], and the next year had the theory of colors, and in May following I had entrance into the inverse method of fluxions [integral calculus]. And the same year I began to think of gravity extending to the orb of the moon, and having found out how to estimate the force which a globe revolving within a sphere presses the surface of the sphere, from Kepler's rule of the periodical times of the planets . . . I deduced that the forces which keep the planets in their orbs must be reciprocally as the squares of the distances from the centers about which they revolve: and thereby compared the force requisite to keep the Moon in her orb with the force of gravity at the surface of the earth, and found them to answer pretty nearly. . . . (Fauvel et al. 1988, 14)

Even allowing for overstatement and for the fact that Newton rarely acknowledged the insights he got from others, this catalog represents an amazing set of achievements.

It took nearly half a century for Newton to work out all of the details of the mathematics, optics, and celestial mechanics grounded in the fundamental insights achieved during this eighteen-month period. By 1669, he had written *De Analysi* on finite and infinite series expansions and on the integration of various mathematical expressions using series approximations. The manuscript was privately circulated but it was not published until 1711. Similarly, he completed his more general work *The Method of Fluxions* in the early 1670s and circulated it among friends; but it was not printed until after his death in 1736. In 1669, Newton lectured on optics in Cambridge, writing out his

Optical Lectures, which were also published after his death in 1728. Finally, in 1672, Newton submitted his first scientific paper on "A New Theory of Light and Colors" to the Royal Society of London. As a result of what Newton considered to be unwarranted criticisms of this paper by Robert Hooke, Newton withheld *Opticks*, the longer work that he was preparing, from publication until after Hooke's death in 1704. His great work on celestial (and terrestrial) mechanics was put on hold until a young friend, Edmund Halley, reported that Hooke was speculating about what kind of paths the planets would follow if they were attracted toward the sun by inverse square forces. Unwilling to be preempted by his nemesis, Newton entered another period of intense work, which resulted in the publication of the *Principia* in 1687.

In the optical lectures of 1669, Newton was already promoting his synthesis of mathematical and experimental methods. Urging mathematicians to study nature and natural philosophers to study mathematics, he argued that if both groups do as he proposes:

> the former shall not entirely spend their time in speculations of no value to human life, nor shall the latter while working assiduously with an absurd method, perpetually fail to reach their goal. But truly with the help of philosophical geometers and of geometrical philosophers, instead of the conjectures and probabilities that are blazoned about everywhere, we shall finally achieve a natural science supported by the greatest evidence.

Specifically with respect to the mathematically formulated propositions on colors presented in the lectures, he argued: "these propositions are to be treated not hypothetically and probably, but by experiments or demonstratively" (Fauvel et al. 1988, 86).

It was in the *Principia* that Newton presented the most spectacular illustration of his new synthesis at work. In the first book of the *Principia*, Newton derived a series of propositions regarding how point masses and aggregate bodies composed of these point masses would move if those bodies were attracted to one another according to inverse square law forces. This book was entirely mathematical, making no claims that any such bodies actually exist.

Book 2 used these propositions to demonstrate that Descartes' natural philosophy could not be correct. Then in book 3, Newton set out to demonstrate that the propositions of book one could, in fact, be used to prove not only that the planets are drawn to the sun by inverse square forces, but that all moons, including that of the earth, are drawn to their planets by inverse square law forces, and, indeed, that every body in the universe is drawn to every other body in the universe by a force proportional to the inverse square of their distances from one another.

The proof that the planets were drawn to the sun by inverse square law forces followed directly from Kepler's observed laws of the motions of the planets, which had been established by 1619. Kepler had shown that the planets do, in fact, move around the sun in elliptical paths, that the radii drawn from the sun to the planets cut out equal areas in equal times, and that the periods of motion of the different planets are proportional to the three-halves power of their mean distances from the sun. Given the mathematical proofs from book 1, it followed that the planets were drawn to the sun by an inverse square law force. To get from the inverse square law force between the sun and planets to the completely universal law of gravitation, however, Newton had to introduce a set of statements that he identified as "Rules of Right Reasoning in Natural Philosophy." Slightly modified from edition to edition, by the third edition of 1726 they appeared as follows:

Rule 1: We are to admit no more causes of natural things than such as are both true and sufficient to explain their appearances.

To this purpose the philosophers say that Nature does nothing in vain, and more is in vain when less will serve; for nature is pleased with simplicity, and affects not the pomp of superfluous causes.

Rule 2: Therefore to the same natural effects we must, as far as possible, assign the same causes.

As to respiration in man and in beast; the descent of stones in Europe and in America; the light of our culinary fire and of the sun; the reflection of light in the earth and in the planets.

Rule 3: The qualities of bodies, which admit neither intension or remission of degrees, and which are found to belong to all bodies within the reach of our experiments, are to be esteemed the universal qualities of all bodies whatsoever. . . .

Rule 4: In experimental philosophy we are to look upon propositions inferred by general induction from phenomena as accurately or very nearly true, notwithstanding any contrary hypotheses that may be imagined, till such time as other phenomena occur, by which they may either be made more accurate, or liable to exceptions. (1962, 397–400)

Combining rules one and two with the observation that the moons of Jupiter and Saturn also move according to Kepler's laws, we find that they must not only be drawn toward their planets by inverse square law forces, but that those forces must be the same forces as those that draw the planets to the sun. Furthermore, even though the earth has only one moon, so that we cannot compare its period with that of any other moon, the earth's moon does move in a nearly elliptical path such that equal areas are cut out in equal times by a line drawn from the center of the earth to the center of the moon. Thus, by rules one and two, the moon must also be pulled to the earth by the same force that pulls the planets to the sun and other moons to their planets.

Furthermore, since all bodies on the surface of the earth fall toward the center of the earth with a force that is just that which would cause the motion of the moon in its orbit if the inverse square law force felt at the moon were acting at the earth's surface, they must be attracted to the earth by the same inverse square law force according to rule one. Finally, because all of the bodies that we experience directly draw one another by the same gravitational force, by rule three, we must hold that every body in the universe draws every other body in the universe according to the same gravitational force.

Newton held that the law of universal gravitation was certain, but this certainty was purchased only by admitting that the cause of gravity was unknown. Rejecting the appeal to mechanical and occult causes alike, he wrote:

> I have not been able to discover the cause of [the] properties of gravity from phenomena, and I frame no hypotheses . . . and hypotheses, whether metaphysical or physical, whether of occult qualities or mechanical, have no place in experimental philosophy. In this philosophy particular propositions are inferred from the phenomena and afterward rendered general by induction. . . . To us it is enough that gravity really does exist and act according to the laws we have explained. . . . (1962, 547)

Rule four of right reasoning guaranteed that his results could not be challenged on the grounds that they were inconsistent with any hypothesis.

Newton's science thus seemed of a substantially different kind both from the mechanical philosophy of persons such as Charleton and Boyle, who made no claims to certainty, and from the Platonic antimaterialist philosophies of Cudworth and Ray, who claimed to be able to demonstrate the necessity of a plastic nature from basic metaphysical or theological assumptions about the goodness and power of God. Using the universal law of gravitation, Newton and Newtonian natural philosophers were able to solve a huge range of outstanding problems in celestial mechanics and natural philosophy, and many eighteenth-century figures hoped to be able to extend his methods and bring the same kind of certainty to their understandings of phenomena of all kinds, including those that we would identify as psychological, political, and religious.

NEWTON AND PROPHECY INTERPRETATION

During the time when Newton was working on the materials that would be brought together into the *Principia* and formulating his "Rules of Right Reasoning in Natural Philosophy," he was also deeply involved in trying to interpret the prophetic books of the Bible. On the one hand, Newton was concerned with the prophetic writings because he believed that they were the most important books of the Bible:

> Giving ear to the prophets is a fundamental character of the true church. . . . [T]he authority of Emperor, King, and Princes is human. The authority of Councils, Synods, Bishops, and Presbyters is human. The authority of the Prophets is divine, and comprehends the sum of religion. . . . Their writings contain the covenant between God and his people, with instructions for keeping this covenant. (Newton 1733, 252)

On the other hand, he seems to have been fascinated with them for a reason that had been articulated most clearly by Robert Boyle, though John Wilkins and John Locke would have agreed fully. This group had argued that the miracles associated with biblical claims were the best evidence to guarantee both the authority of the Bible and the authority of Christ as portrayed in the Bible. But of all the kinds of miracles, fulfilled prophecies were the most convincing because they continued into the present and thus could be directly confirmed, without the mediation of witnesses from long ago, whose testimony might be questioned. In Boyle's words,

> True prophecies of unlikely events, fulfilled by unlikely means, are supernatural things; and as such, (especially their author and design considered) may properly enough be reckoned among miracles. And, I may add, that these have a peculiar advantage above most other miracles, on the score of

their duration: since the manifest proofs of the prediction continue still. (1744, 5:526)

If one could determine unambiguously the meaning of the Bible's prophecies, especially of events in the present, then one would have a way to test the authenticity of the biblical miracles directly.

Just as Newton formulated rules for natural philosophizing, he formulated rules for interpreting the Bible that were formally drafted variously as "Rules for interpreting the words and language in Scripture," "Rules for Methodizing the Apocalypse," or simply "General Rules of Interpretation." As he proceeded with both endeavors, it seems clear that his methodological ideas in the two domains—religious and scientific—reinforced one another and that they depended strongly on his conceptions of God and of the relationship between God and the creation.

One of the most interesting parallels between his rules for interpreting nature and those for interpreting Scripture is between his first "Rule of Right Reasoning in Natural Philosophy" and the ninth of his "Rules for Methodizing the Apocalypse," which were both formulated at approximately the same time. The first rule in the *Principia* insists that we must admit no more causes than are "true and sufficient" to account for phenomena because "nature is pleased with simplicity, and affects not the pomp of superfluous causes." The ninth rule for prophecy interpretation insists upon constraining the interpretation of scriptural passages based on the principle of simplicity:

[One must] choose those constructions [i.e., interpretations] which, without straining, reduce things to the greatest simplicity. . . . Truth is ever to be found in simplicity, and not in the multiplicity and confusion of things. As the world, which to the naked eye exhibits the greatest variety of objects, appears very simple in its internal constitution when surveyed by the philosophic understanding, so it is in these visions. It is the perfection of all God's works that they are done with the greatest simplicity. He is the God of order and not confusion. And therefore as they that would understand the frame of the world must endeavour to reduce their knowledge to all possible simplicity, so must it be in seeking to understand these visions. (Manuel 1974, 120)

In claiming that God works in the simplest possible way, Newton was actually rejecting an important aspect of traditional natural theology, which praised God's creativity that was revealed in the great variety of ways in which he designed living beings to accomplish the same functions. Charleton, for example, had pointed out that God used feathers on birds but skin on bat wings to accomplish the same lift.

A critical consequence of Newton's confidence that God worked both in Scripture and nature in the simplest possible way was that this assumption offered a strategy that was immensely important for both his science and his religion. Without it, he could not have arrived at the law of universal gravitation, and he would have had no way to argue for a single "best" reading of the prophecies.

Another principle established by Newton in interpreting the prophetic books also paralleled a section of the first rule in the *Principia*. In rule one, the claim that a cause must be "true" as well as sufficient meant that it must have been demonstrated as operating in some context other than the one currently under investigation. So, when talking about gravity as the cause of the fall of bodies on the surface of the earth, Newton argued that it had already been established as "true" in connection with the motions of the planets and their moons. By the same token, Newton argued that we must establish the "true" use of the language used in the prophecies of Daniel and John by going outside of those texts. Following hints from both Richard Hooker and Joseph Mede, a widely read early-seventeenth-century biblical scholar, Newton argued that one must establish that the terms used by the prophets had an unambiguous and well-established meaning within the Near Eastern culture that produced those texts. Thus, he sought to establish what he called a "hieroglyphic dictionary" or a dictionary of the "prophetic style" drawn from the prophetic writings of the tribes and nations surrounding the Hebrews and early Christians. Any interpretation of scriptural prophecy would then have to be consistent with the linguistic usages of the prophetic style. Thus, Newton wrote,

> if any man interpret a beast to signify some great vice, this is to be rejected as his private imagination because according to the style and tenor [of the prophetic style] . . . a beast signifies a body politique and sometimes a single person who heads that body, and there is no ground in Scripture for any other interpretation. (Manuel 1974, 120)

In the hands of later higher critics of the Bible, this suggestion, promoted in the writings of William Whiston, Newton's closest religious disciple, became the foundation for "form criticism," which generally argued that the meaning of biblical passages could be discovered by investigating similar passages in the literature of nearby societies where the meaning could be discovered more easily. As in the case of rules one and nine above, I know of no way to discover whether Newton's emphasis on an independently established "true" cause in nature suggested the need for an independently established "true" linguistic usage, or vice versa.

In one case it does seem clear that Newton developed a methodological rule in connection with prophecy interpretation and only subsequently applied it to natural philosophy because the religious version can be dated to 1680 while the natural philosophy version does not appear until the fourth (1726) edition of the *Principia*. Rule eleven of the "Rules for methodizing the Apocalypse" is the final rule and claims that any interpretation arrived at using the preceding rules should be accepted even if alternative readings are offered, because all other interpretations must be "grounded upon weaker reasons [and] . . . that is demonstration enough that it is false" (Manuel 1974, 121). Similarly, Rule four of reasoning in the fourth edition of *Principia* is the final rule of the "Rule of Right Reasoning in Natural Philosophy" and states that propositions induced using the first three rules should be accepted in the absence of new information, not withstanding alternative hypotheses. It concludes: "This rule we must follow that the argument of induction may not be evaded by hypotheses" (Newton 1962, 400).

Newton did not make his views on prophecy interpretation public during his lifetime, probably in part because he feared that they would not be well-received and would damage his reputation. Instead, he urged Whiston to try them out in a series of works. These works caused Whiston to lose the Lucasian Professorship, confirming Newton's judgment.

NEWTONIAN RELIGION

Among those who sought to use Newton's scientific work for religious purposes, one man, the Cambridge mathematician John Craig, deserves special mention, not because of his long-term impact but because his work serves as an illustration of how Newton's concerns with the evidentiary foundations of Christianity were sometimes carried to extremes by his followers. In 1699, Craig published *Theologia Christiane Principia Mathematica* (*Mathematical Principles of Christian Theology*). Organized to parallel Newton's *Principia*, this odd work developed formal rules of historical evidence and probability in order first to calculate the degree of confidence warranted in the truth of the Gospels and then to establish the outer limit of time for the second coming of Christ. Craig argues that one should not believe in the Gospels without evidence. That evidence comes from the witnesses to Christ's miracles, and our confidence in the testimony of witnesses decreases over time. Somehow he established that in 1699 the evidence in the truth of the gospel narrative was equal to that of the uncontested statements of twenty-eight disciples. Because of the decay in confidence, however, he argued that by the year 3144 the confidence level would decrease below that corresponding to half of one witness; so there would be less reason to believe than to doubt. As a consequence, the second coming must occur before 3144, while there is reason to retain at least some faith in Scripture.

The general consensus among eighteenth-century Anglican apologists was that Craig's work was ill-advised and poorly executed, and that his work was "altogether as damaging to the Christianity as the skepticism he thought he was combating" (Craig 1964, 1). Most attempts to use Newtonian science for religious purposes elicited much more positive response, at least initially.

Newton was moderately interested in natural theology, but in the first editions of the *Principia* and *Opticks* he made very few allusions to their use for religious purposes. He was, however, contacted in 1691 by Richard Bentley, who requested evidence from Newton's works that could be used to undermine atheism. Bentley had been chosen by the executors of Robert Boyle's will to give the first annual series of "Boyle Lectures" endowed under the terms of the will and aimed at promoting "the truth of the Christian Religion in General, without Descending to the Subdivisions Among Christians" (Olson 1990, 111). The request led Newton to write a series of letters to Bentley and to add the religiously oriented "Scholia" to the *Principia* and "Queries" to the *Opticks*.

Newton told Bentley that it was enough to understand the first few and simplest propositions of books 1 and 3 of the *Principia* in order to understand its most important theological consequences. Even better, from Bentley's perspective, Newton wrote a series of four letters to him, which Bentley followed closely in *A Confutation of Atheism from the Origin and Frame of the World, The Seventh Lecture*, published in 1692.

Newton was inclined to agree with the corpuscular philosophers, including Boyle, that matter is essentially passive, for that ensured the need for God to somehow initiate and conserve its motions. But the universal law of gravity demonstrated that every bit of matter was attracted to every other bit. As a consequence, Newton suggested that there must be some cause of gravity that is not intrinsic to matter, possibly either God or some immaterial agent of God. Bentley concurred, concluding that gravity is "above all mechanism and material causes and proceeds from a higher principle, a Divine energy and impression" (Cohen 1958, 344). A few years later, another Boyle lecturer, William Whiston, was even more emphatic on this issue, writing, "Tis now evident that Gravity . . . depends entirely on the constant and efficacious, and, if you will, the supernatural and miraculous Influence of Almighty God" (1708, 284). Yet another Newtonian Boyle lecturer, Samuel Clarke, was less inclined to see gravity as caused by the direct action of God. He preferred to think of it as produced by some subordinate instrument like the Cambridge Platonist's plastic nature (1706, 19–23). Regardless of whether they sought recourse to God or to some immaterial agent of God, Newtonian natural theologians were generally agreed that gravity provided irrefutable evidence against atheism.

Proposition ten of book 3 of the *Principia* raised a second major issue for Bentley because it suggested that, while the solar system might be stable if each planet interacted only with the sun, the interactions among the planets would produce a tendency for the whole system to collapse over time. Bentley argued that, under these circumstances, the direct providential activity of God was necessary to stave off collapse (Cohen 1958, 349–350). Newton concurred and included "Query 23" to the 1706 edition of his *Opticks* in which he said that even if the system had not yet required a special providential adjustment, it would do so eventually due to "inconsiderable irregularities . . . which may have arisen from the mutual actions of comets and planets upon one another, and which will be apt to increase, till this system wants reformation" (1952, 402). Colin McLaurin, whose *An Account of Sir Isaac Newton's Philosophical Discoveries* of 1748 made the arguments of the *Principia* accessible to a much larger audience, also agreed, writing, "the Deity has formed the Universe dependent upon himself, so as to require to be altered by him, though at very distant periods of time" (Odom 1966, 542).

The responses to the Newtonian argument for the existence of a specially providential God grounded in the apparent imperfection of the universe illuminates the extent to which confidence in God's providential care for humans was pervasive in seventeenth-century European culture. No one, to my knowledge, responded by worrying about the imminent collapse of the world, as many would probably do today if Stephen Hawking or Kip Thorne were to declare that their cosmological theories predicted that the world would implode in a short period of time. Instead, they either joined the Newtonians and reveled in the fact that there was now evidence from nature for the need for a specially providential God, or they joined with Gottfried Leibniz, the German philosopher and co-inventor of calculus, in arguing that the focus on imperfection was so ridiculous that it should undermine confidence in Newton's natural philosophy.

Leibniz was offended by the fact that the Newtonians implied that God was so inept that he could not create a lasting universe, and he wrote to Princess Caroline of Wales, the heir-presumptive to the English Crown, complaining that Newtonianism was a central cause of the decline of morality in England. Caroline brought this letter to the attention of Samuel Clarke, and Clarke responded in a way that surely must have endeared Newtonianism to the royal princess:

> As those men, who pretend that in an earthly government things may go on perfectly well without the king himself ordering or disposing of anything, may reasonably be suspected that they would very well like to set the king aside, . . . so too those who think that the universe does not constantly need God's actual government, but that the laws of mechanism alone would allow phenomena to continue, in effect tend to exclude God out of the World. (Alexander 1956, 14)

When Bentley chose to appeal to Newton's natural philosophy in order to "confute atheism" in his Boyle Lectures of 1691, he initiated a pattern of religious appeals to Newtonian science that was followed by Whiston in *A New Theory of the Earth, From its Original to the Consummation of All Things, Where the Creation of the World in Six Days, The Universal Deluge, and the General Conflagration, As Laid Down in the Holy*

Scriptures, Are Shewn to be Perfectly Agreeable to Reason and Philosophy (1696); by Samuel Clarke in *Demonstration of the Being and Attributes of God* and in his *A Discourse Concerning the Unchangeable Obligations of Natural Religion, and the Truth and Certainty of the Christian Revelation* (1706); as well as by the most popular of the Boyle lecturers, William Derham, who wrote both *Physico-Theology: or, A Demonstration of the Being and Attributes of God, from the Works of Creation* (1713) and *Astro-Theology: or a Demonstration of the Being and Attributes of God from a Survey of the Heavens* (1715).

Unfortunately for those who accepted the Newtonian argument from imperfection and saw in it a key evidence for a specially provident God, subsequent developments in Newtonian celestial mechanics during the late eighteenth century showed that Newton's concerns about the collapse of the solar system were misplaced. Newton had approximated mathematical expressions by using the first few terms of their infinite series representations in arriving at his conclusions about the instability of the solar system. The eighteenth-century French mathematicians Jean D'Alembert, Louis Lagrange, and Pierre Simon Laplace, however, showed that Newton's approximations were sufficiently inexact that collapse was far from certain. When Napoleon Bonaparte asked Laplace where God was in *System de la Monde,* Laplace is reported to have answered: "Sir, I have no need of that hypothesis" (Odom 1966, 535). As the eighteenth century wore on, then, it was the Deists who increasingly drew comfort from Newtonian natural philosophy, because they could argue that the design argument established God as creator but failed to offer evidence for a God continually engaged with his creation.

JOHN LOCKE AND THE RISE OF DEISM

The emphasis on natural theology that developed among seventeenth-century Anglicans, and which culminated in the various Boyle Lectures, continued through the eighteenth century and into the nineteenth century. Bishop Joseph Butler contributed a fascinating argument that claimed that even the doctrine of life after death was supported by nature in the metamorphosis of caterpillars into butterflies in *The Analogy of Religion* (1736). William Paley produced what was probably the most comprehensive example of the genre in 1802 when he published *Natural Theology: Evidences of the Existence and Attributes of the Deity, Collected from the Appearances of Nature,* and in the 1830s the eighth Earl of Bridgewater even tried to emulate the earlier Boyle lectures by bequeathing money to endow a series of eight Bridgewater Treatises intended to appropriate the latest developments in the various sciences to Christian ends. From the standpoint of many Christians, however, one of the most important—and disturbing—consequences of the vogue for natural theologizing initiated among seventeenth-century liberal Anglicans was that it seemed to undermine the importance of Scripture-based religion and to lead many into believing that natural religion was not so much a support to Christianity as a superior alternative to it.

William Whiston already recognized this unintended and troublesome trend as early as 1717 in discussing the implications of Newtonian natural theology in particular. By virtue of its very success, natural theology seemed to him to leave only revealed religion as a target for those who were irreligious out of ignorance, perversity, or madness:

> [C]ertain persons not overly-religiously disposed [were] soberly asked after Dr. Bentley's remarkable Sermons at Mr. Boyle's lectures, Built upon Sir Isaac Newton's Discoveries, and leveled against the prevailing Atheism of the Age, What had they to say on their own Vindication against the evidence produced by Dr. Bentley? The answer was, That truly they did not well know what to say against it, upon the head of Atheism: But what, they say, is this, to *the fable of Jesus Christ*? And in confirmation of this Account, it may, I believe, be justly observed, that the present gross *Deism*, or the opposition that has of late so evidently and barefacedly appeared against Holy Scripture, has taken its date in some measure from that time. (Force 1985, 45–46)

Though Deism, or the belief that the world must have been created by an intelligent designer but that he has not subsequently been involved in its processes, had already been formulated in the sixteenth century, Whiston was probably right in dating its broad popularity among intellectuals to the beginning of the eighteenth century. What he failed to acknowledge, however, was the importance of the works of John Locke (1632–1704) in encouraging this development.

Locke had certainly not intended to encourage this trend. In fact, in *Essay Concerning Human Understanding* in 1690, *The reasonableness of Christinity* in 1695, and *Discourse of Miracles* in 1701, Locke was clearly trying to develop a theory of knowledge grounded in contemporary science that could be used to show that the evidence for Christianity was every bit as good as the evidence for natural philosophical knowledge. But as he grappled with the problems of evaluating the evidence for revealed religion and the problems of analyzing just what miracles are—that they might authorize religious beliefs—he often inadvertently did more to undermine belief in revealed religion than to assure people of its validity. In *Essay Concerning Human Understanding*, for example, Locke sought to demonstrate that humans could only understand a concept if they possessed its elements through sensory experience or reflection on the processing of that experience. As a consequence, he had to conclude that scriptural revelation could not add any understanding beyond what we have from ordinary experience and thought. Similarly, he argued that the language of the Bible offers immense problems of interpretation.

There were, however, certain claims of revealed religion that Locke tried to evaluate in the light of natural knowledge. One of these was the immortality of the soul and the consequent possibility of life after death. Locke demonstrates in book IV of the *Essay* that we cannot have certain knowledge of the coexistence of properties in substances. Thus, though we may have well-founded ideas of matter, thought, and volition, we cannot know how or whether the three are linked to one another. In particular, we cannot know whether that part of us that we experience as thinking and willing, and which Christians call the soul, is material or immaterial (Locke 1959, 2:192–193).

For Locke, it mattered little whether there was an immaterial soul or whether matter itself could think. For most Christians, however, this was a tremendously important issue because the immateriality of the soul seemed to be a precondition for its immortality, and the immortality of the soul was critical if there were to be punishments and rewards after death. That is, the whole theory of salvation and damnation seemed to hinge on the immateriality of the soul and to bring us back to the need for revelation to complete the foundations of Christianity. This need in turn led to the question of why we should believe in the authority of Scripture, supposing that we can figure out what it means. Turning to what evidence there was to authorize our belief in revealed religion, Locke ultimately concluded, with Boyle and others, that revelations can only be guaranteed by the miracles that accompany them: "These supernatural signs [are] the only means God is conceived to have to satisfy men as rational creatures of the certainty of anything he would reveal as coming from himself" (Locke 1958, 85).

When he wrote *Essay Concerning Human Understanding*, Locke apparently believed that miracles were completely self-confirming and non-problematic, but hostile and friendly critics alike soon showed him that this was not so. How can we recognize when a miracle has occurred? Locke attempted to answer this question in *A Discourse on Miracles*, which was undoubtedly pious in intent. But the work had an impact that would have shocked its author if he had been alive to recognize it, because it seemed to do more to legitimize anti-scriptural Deism than any other single Christian text.

According to Locke, miracles are sensible operations that are beyond the comprehension of those who witness or hear about them. Thus, they are supposed by humans to be contrary to the ordinary course of nature and to be of divine origin. They may, however, just be part of a more complex natural process than some, or even all, humans are capable of understanding. There is nothing in Locke's doctrine to even faintly hint at the extraordinary suspension of law that for Charleton, Boyle, and Newton allowed God's absolute power and the political absolutism of the monarch to mutually explain one another. It now appeared as though God and monarch alike governed solely by laws, some of which we could—just partially—grasp.

Furthermore, one might ask how miracles could guarantee the authenticity of Scripture if different persons have greater or lesser understandings of nature, so that what is miraculous to some observers is not to others (Locke 1958, 80). Locke dealt with this question by insisting that there are such demonstrable limits to human understanding that certain events must be miraculous for all possible witnesses; and only these events can be genuine signs of God's authority. But he was still faced with the problem of identifying such signs; for the simple fact that nobody presently understood a phenomenon was no guarantee that they never could. Once more Locke was forced back to a doctrine of probabilism. Something was probably a genuine miracle if, firstly, it evinced supreme power. Thus, for example, the fact that Moses' serpent ate the serpents of the Egyptian priests provides evidence in favor of Moses' serpent being miraculous while those of the Egyptians were not (Locke 1958, 83–85). Secondly, miracles were to be judged more probable if many were associated with a single individual, such as Christ (Locke 1958, 83). Thirdly, miracles were to be judged more probable if they were associated with messages that could not have been known without revelation. Finally, since God cannot be inconsistent, we cannot accept as a divine miracle any extraordinary event associated with a message that is inconsistent with natural religion and natural morality (Locke 1958, 84).

Almost no one found Locke's theory for the identification of genuine miracles compelling. So, as a result, his final work undermined virtually all claims that there was convincing evidence for the authority of Scripture to be found in rational analysis. If one took Locke seriously, then one was left with precisely the options that he and his liberal Anglican friends from Hooker on had been trying to avoid. One could accept the authority of Scripture on the basis of an inner light or a religious sense granted directly from God, as many dissenting inheritors of the Puritan tradition did. One could acquiesce to the authority of tradition, as the Catholics and now the High-Church Anglicans chose. Alternatively one could admit that there were no adequate grounds for belief in Scripture-based religion, as the new crop of Deists and Free Thinkers claimed.

In England, all of these options were entered into by significant numbers of intellectual leaders and followers. On the continent, however, especially in France, the response of intellectuals during the early and middle years of the eighteenth century was overwhelmingly deistic or atheistic. There, a bitter anti-clerical mood existed among intellectuals. This mood was particularly well-articulated by Claude-Adrien Helvetius in *De l'Esprit* in 1757. Railing against the leaders of Christian sects who seemed to promote more hate than love in the course of their efforts to extend or protect their own power, he wrote:

> If we cast our eyes to the North, the South, the East, and the West, we everywhere see the sacred knife of religion held up to the breasts of women, children, and old men; the earth smoking with the blood of victims sacrificed to the false Gods or to the Supreme Being; every place offers nothing to the sight but the vast, the horrible, carnage caused by a want of toleration. (Olson 1990, 267–268)

Helvetius followed his anti-clericalism into atheistic materialism, as did his friend, the Baron D'Holbach; but both Voltaire and Jean-Jacques Rousseau followed theirs into a more common emphasis on natural religion.

Throughout his life, Voltaire promoted religious toleration and ridiculed the differences among Christian denominations. His antipathy to organized religion became an obsession after the 1762 Calas affair, in which a Protestant father was convicted, without evidence, of killing his son for planning to convert to Catholicism. After this episode, Voltaire wrote pamphlets and books in an effort to undermine the power of the official Gallican Catholic Church, and he sought to replace Christian doctrines with a natural religion:

When reason, freed from its chains, will teach people that there is only one God, that this God is the universal father of all men, who are brothers; that these brothers must be good and just to one another, and that they must practice all the virtues; that God, being good and just, must reward virtue and punish crimes; surely, my brethren, men will be better for it, and less superstitious. (Brooke 1991, 163–164)

Though Rousseau was less of an activist in criticizing organized religion, he expressed his support of natural religion in a segment, "The Profession of Faith of a Savoyard Vicar," of his educational novel, *Emile* (1762). Rejecting such Christian doctrines as original sin and redemption through Christ, Rousseau offered a natural religion based primarily on portions of Samuel Clarke's 1705 Boyle lectures on natural religion.

In Britain, the most able early eighteenth-century promoters of Deism and free thought were Matthew Tindal and Anthony Collins. Both men pushed Locke's ideas regarding the empirical origins of all human knowledge into a denial of the doctrine of the Trinity and into a rejection of Christian "mysteries" in general, and Collins drew heavily from the Newtonian natural theologians as well. Even some members of the Anglican hierarchy seemed infected by the Lockean demand that everything for which no clear idea could be discovered must be eliminated from the beliefs demanded of a Christian. Thus, Bishop Gilbert Burnet argued not only that the Church should tolerate the denial of the doctrine of the Trinity, but that it should deny the existence of any of the seven Christian "mysteries" mentioned in the Bible:

It is a question, whether those who plead for Mysteries can believe themselves, after all their zeal for them: since a man can no more think that is true of which he has no Idea than a man can see in the dark; for let him affirm ever so much that he sees, all other persons who perceive it to be dark are sure that he sees nothing. (Sullivan 1982, 96)

It was only the smallest step from this position to that of the Deist Tindal, who argued that there could be nothing true about Christian revelation that was not independently present in natural religion.

REACTIONS AGAINST NEWTONIAN NATURAL THEOLOGY

Those who were deeply disturbed by the ways in which Latitudinarian and Newtonian natural theologies seemed to divert attention from revealed religion, to rob religion of much of its emotional appeal, and even to encourage heterodox beliefs such as Deism and Unitarianism, responded in a variety of ways. Just as twentieth-century creation scientists have tried to offer a biblically based alternative to the Darwinian theory of evolution by natural selection because they find the implications of that theory unacceptable, a substantial group of High Church Anglicans and evangelicals developed and promoted a biblically based natural philosophy. This strategy was first articulated by John Hutchinson (1634–1737) in *Moses' Principia* in 1724 and explored in Samuel Pike's *Philosophia Sacra: or the Principles of Natural Philosophy Extracted from Divine Revelation* of 1753. Among the many advocates of Hutchinsonian natural philosophy were such High Church Anglicans as President Samuel Johnson of King's College in New York, and George Horne, who became Bishop of Norwich. The Scottish conservative Presbyterian Duncan Forbes spread the doctrine in Edinburgh, where it was taken up by several major scientists, including James Hutton, while John Wesley promoted Hutchinsonian natural philosophy among evangelicals.

Hutchinson's natural philosophy, rather than challenging the doctrine of the Trinity, embraced it. His theology, rather than demanding a literal reading of Scripture, encouraged an allegorical reading. Rather than working from the King James Bible, Hutchinson insisted on working from the Hebrew version, in which words were often open to a variety of interpretations. For example, in Genesis 1:2, the English Bible describes the earth as "without form and void," implying the existence of empty space. Hutchinson, however, argues that the Hebrew word translated as void, really means "to move or yield easily." As a consequence, Hutchinson argues against empty space and in favor of a space "full of such Matter, as would shift upon the Approach of any other Matter and let it take its place" (Wilde 1980, 3). When he comes to discuss the material foundation of all things, he argues that fire, light, and air, being the analogs of the three persons of the Trinity, constitute all things and that, like the Father, Son, and Holy Ghost, which are all modifications of God, they are merely three different modifications of single universal ether (1980, 4). Finally, because he is concerned with maintaining the traditional distinction between passive matter and active spirit, Hutchinson goes back to the earlier mechanical philosopher's argument that the three forms of matter must have their activity from God and that they can only interact by contact action. One can see, then, that Hutchinson and his followers reversed the pattern of natural theology, which sought to discover God's attributes from nature. Conversely, they sought to discover the attributes of nature in Scripture.

One of the most important critiques of natural theology—because it stimulated both German philosopher Immanuel Kant's completely new way of thinking about the relationships between science and religion and Charles Darwin's theory of evolution by natural selection—came from within the Newtonian tradition itself, at the hands of a Scottish philosopher, David Hume (1711–1776). Hume's entire career was devoted to extending the Newtonian method of philosophy to a huge range of topics. When he attempted to do that to religion, he came up with some very surprising and disturbing results.

In *Dialogues Concerning Natural Religion* (1779), which was not published until after his death, though it had been written more than a quarter of a century earlier, Hume turned his critical skills on the design argument that had been such a central feature of Anglican natural theologizing in the seventeenth century, and which would continue to be powerful in spite of his cogent remarks. Hume argued that even if we accept that the world evidences design, that evidence does not assure us that there is a single rather than a number of designers. But, he argues, there are apparent sources of order other than design. While artificial objects might be designed, natural objects need not be. The universe might be more like an animal or a vegetable than like a watch, for example, dependent on a principle of order that is like whatever internal vital principle directs the growth of an organism. This, of course, had been the argument of the Renaissance neo-Platonists.

But Hume offered an even more disconcerting argument that denied the necessity of any goal-oriented development at all. Hume suggested that the particles that constitute the universe are constantly moving—supposedly in a random fashion. Given a long enough time, they must throw together groups of particles in every possible configuration, and some of those configurations may be at least temporarily stable—that is, once they have been formed they might interlock with one another so that they continue to exist over large periods of time. Indeed, over time, every imaginable stable configuration of atoms must come into existence and eventually decay, as animals do. Each may appear to have been designed for some purpose; but, in fact, no purpose was at work in creating them. Given enough time, some of the configurations produced will seem purposive.

Suppose now that the world were even thrown into a more highly organized form than our own. Hume argues, "Must it not dissolve as well as the animal, and pass through new positions And situations; till in a great, but finite succession, it fall into the present, or some such order?" (1993, 87). That is, if we allow enough time in a universe of randomly moving particles, the configuration that the world presently has will inevitably be produced, without any organizing principle whatsoever. And that is true whether the present order is approached from a condition of less order or from a position of greater order. Charles Darwin was among the first persons to take this argument seriously, though it is alive and well in the form of the infinite universes hypothesis associated with quantum mechanics at the beginning of the twenty-first century. In the current form, it is posited that there are an infinite number of possible universes, of which one universe had to have exactly the characteristics of that in which we live, and we just happen to be living in that one.

Consideration of attempts to establish the existence and character of God through natural theology thus left Hume in a skeptical quandary. There seemed no way to give any significantly greater degree of probability to the existence of a single god, multiple gods, or no god at all using the techniques of traditional natural theology.

Other conservative critics of Newtonian natural philosophy, such as George Berkeley, the Bishop of Cloyne, sought to argue that the vaunted "reason" by which Newtonians were claiming that we should judge Scripture, was badly used by the Newtonians themselves. As one of the outstanding early works usually relegated to the philosophy of mathematics, Berkeley's *The Analyst* (1753) demonstrated that there were logical errors in the way that Newton formulated his calculus—in one step of the argument, certain quantities were assumed to be small but finite, while in another step of the same argument they were allowed to become zero. Turning this claim against the opponents of Christian mysteries, such as the Trinity, Berkeley wrote:

He who can digest a second or third fluxion [i.e., derivative] . . . need not, methinks, be squeamish about any point in Divinity. . . . But with what appearance of reason shall any man presume to say that mysteries may not be the objects of faith, at the same time that he, himself, admits such obscure mysteries to be the objects of science. (Fauvel et al. 1988, 214)

An even greater number turned against Newton's natural theology, though they continued to praise Newtonian natural philosophy. Such, for example, was John Arbuthnot, an advocate of Newton's natural philosophy, who led the group within the Royal Society that supported Newton's claim to the invention of the calculus against the advocates of Leibniz's claims. Arbuthnot, however, was a high churchman who was far more interested in Scripture and far less confident regarding the abilities of human reason to understand the workings of God than the natural theologians, who he viewed as excessively prideful.

A similar sentiment was often expressed by the famous author Samuel Johnson, who went a step further, arguing that natural philosophy, and not merely natural theology, was potentially troublesome because it diverted human attention from more important moral issues. In *Life of Milton*, published in 1779, he criticized Milton for including natural philosophy in the curriculum of a school that Milton ran in his youth. Appealing to Socrates, who he admired, Johnson wrote:

It was his [Socrates'] labour to turn philosophy from the study of nature to speculations on life; but the innovators whom I oppose are turning attention from life to nature. They seem to think that we are placed here to watch the growth of plants, or the motions of the stars. Socrates was rather of the opinion, that what we had to learn was how to do good and avoid evil. (Olson 1983, 194)

For the most extreme attacks upon natural theology, however, one must turn to Jonathan Swift (1667–1715), Anglican Dean of Armaugh, who was a close friend to both Johnson and Arbuthnot. In *Gulliver's Travels,* Swift takes aim at John Ray's discussion of God's fine workmanship evidenced in the microscopic observation of such unlikely objects as lice and flies. Gulliver describes his view of the giant Brobdingnagians and their creatures:

[T]he most hateful sight of all was the lice crawling on their clothes; I could see distinctly, the limbs of those Vermin with my naked Eye, much better than those of a European Louse through a Microscope: and their Snouts, with which they rooted like Swine. They were the first I had ever beheld; and I should have been curious enough to dissect one of them, If I had proper instruments . . . although indeed the Sight was so nauseous, that it perfectly turned my stomach. (Olson 1983, 190)

And in *The Tale of a Tub*, he offers a parody of Boyle's likening the universe to a complex clock and seeing God as the clockmaker. Swift offers us a sect that worships a tailor and sees the universe and everything in it as a magnificent suit of clothes. Within this clothing universe, the human conscience is described as "a pair of breeches; which, though a cover for lewdness as well as nastiness, is easily slipped down for the service of both" (Olson 1983, 188).

For Swift, the opposition to natural theology sometimes slipped into a concern with natural philosophy as well, but the most powerful antagonism to both came from the evangelically oriented mystic and Romantic poet William Blake, whose entire corpus is a sustained lament over what he deemed the destructive force of Newtonian and Lockean ideas, especially the natural theology and natural religion grounded in their works. Beginning with *There Is No Natural Religion* in 1788, Blake's antagonism reached a peak in *Jerusalem*, published in 1804. In this allegory, Albion stands for England:

> [The sons of Albion] build a stupendous building on the plains of Salisbury,
> with chains of rocks round London Stone, of reasonings, of unhewn Demonstrations. . . .
> The Building is Natural Religion and its Altars Natural Morality,
> A building of eternal death, whose proportions are eternal despair. . . .
> Oh Divine Spirit, sustain me on thy wings,
> that I may wake Albion from his long and cold repose;
> For Bacon and Newton, sheath'd in dismal steel, their terrors hang
> Like iron scourges over Albion, Reasonings, like vast Serpents
> Infold around my limbs, bruising my minute articulations.
> I turn my Eyes to the Schools and universities of Europe
> And there behold the Loom of Locke, whose woof rages dire,
> Washed by the Water-wheels of Newton: black the cloth
> In heavy wreaths folds over every Nation: Cruel Works
> Of many wheels I view, wheel without wheel, with cogs tyrannic
> Moving by compulsion each other, not as those in Eden, which
> Wheel within wheel, in freedom revolve in harmony and peace.
> (Olson 1990, 362–363)

These conservative and evangelical attacks did not stop natural theologizing in Britain; but they did signal the growth of an oppositional strain of religion that was to become increasingly important throughout the nineteenth century, especially on the Continent. This strain saw religion as serving emotional needs that could not be touched by the kind of rational approach that was central to natural theology.

Chapter 6

Scientific Understandings of Religion and Religious Understandings of Science, 1700–1859

Among the critical challenges offered to natural theology in the mid- and late eighteenth century, there were three that emerged in the nineteenth century as particularly important. One began with the revival of an ancient tradition of attempts to understand and often to discredit the worship of polytheistic religions. One of the most important central claims of this tradition was that gods were posited, or created, by early humans in an effort to understand dramatic and uncontrollable natural forces. According to the most popular forms of this tradition, to enhance their own power in society, a priestly class exploited the psychological need that humans have for some sense of control over their lives. As the sciences developed an increasing ability to understand and control nature, religion became not only unnecessary, but also a positive impediment to human progress, because the priestly class fought against scientific progress in order to to retain its authority and power. At the hands of Ludwig Feuerbach in Germany this "anthropological" approach to religion was transformed in the nineteenth century to refocus Christianity on attempts to meet current human needs. Moreover, Auguste Comte and his followers in France and elsewhere attempted to use it to replace Christianity with a scientifically grounded "Religion of Humanity." This tradition continues to offer the basic orientation for most modern interpretations of science and religion interactions that insist that "conflict" characterizes their most frequent and important interaction.

In contrast to the revived ancient anthropological tradition, which saw science emerge gradually as a superior replacement for religion, the second set of challenges to natural theology was initiated only at the end of the eighteenth century by the German philosopher Immanuel Kant. Kant developed an interpretation of science that challenged key assumptions about the possibility of any "positive" or "objective" knowledge of the universe that is independent of the human knower. In addition, he developed an interpretation of religion that radically divorced its basic assumptions from those of the sciences. As a consequence, argued Kant, science and religion can neither be in conflict, nor can they support one another. They simply have no ground in common. Variants of the Kantian tradition continue to inform most modern interpretations of science and religion that argue that science is grounded in reason, that religion is grounded in faith, and that reason and faith can have no bearing on one another, either positive or negative.

The third challenge to natural theology came from a version of the ancient anthropological tradition that reinforced the Kantian claim that religious goals were completely divorced from knowledge of the natural world. But this tradition went on to argue that religion instead addressed human feelings of dependance, awe, wonder, isolation, and so forth. From this point of view, most fully developed by the German theologian Friedrich Schleiermacher and his followers, religion is intended to respond to and evoke emotions and sometimes to motivate actions, but not to offer explanations of phenomena. Moreover, religious claims can be neither supported nor undermined by natural scientific knowledge. Within the anthropological tradition that culminated in Schleiermacher's theology, new techniques of interpretation, labeled "hermaneutics," were developed to discern the meanings of biblical texts. This tradition dominated avant-garde nineteenth-century continental Protestant theology and became the foundation for almost all modern versions of "liberal" Christianity.

In response to the anthropological traditions and the Kantian tradition that challenged natural theology and emphasized that science and religion had different goals and different methods, a group of Scottish intellectuals developed a philosophy that sought to reclaim science as a support for religion. This "Common Sense philosophy," which emerged in Scotland in the eighteenth century, became especially important in America during the first half of the nineteenth century. It revitalized natural theology and encouraged the notion that science should be pursued primarily as a religious vocation. Though it provided powerful encouragement for mainstream scientific activity in the nineteenth century, its persistence through the twentieth century has offered the major intellectual support for such deviations from the scientific norm as creation science.

In this chapter, we will consider the development of the anthropological and the Kantian challenges to natural theology as well as the attempt to reconstitute science as a support for religion prior to the appearance of Charles Darwin's *Origin of Species* (1859). Darwin's work signaled the beginning of a new era of considerations regarding science and religion interactions. Neither anthropological, Kantian, or Common Sense arguments were abandoned; but they were transformed in the light of evolutionary developments.

EARLY ANTHROPOLOGICAL APPROACHES TO RELIGION

A special renewed interest in the origins and meanings of "pagan" religious practices and myths was one of the significant outgrowths of the combination of the voyages of discovery and the European colonization of much of the rest of the world, the recovery of Greco-Roman literature by humanists, and the rise of printing based on moveable type technology. This new interest in antiquity arose largely because Europeans became fascinated with the religious practices of the "savage" peoples they were encountering, and the classical analyses of pagan religions offered a framework for understanding those practices.

One approach to pagan mythology not widely practiced in the eighteenth century was the allegorizing tradition. According to this approach, the simple oral myths and texts of early religions were intended as allegories that masked hidden and profound wisdom regarding spirituality and morality. This approach to paganism had predominated during the Renaissance; but in the more matter-of-fact environment at the end of the seventeenth century, it was pilloried by the clever and sardonic Pierre Bayle in his *Dictionaire Historique et Critique (Historical and Critical Dictionary)* of 1697, never to recover its previous authority and popularity. It was additionally discredited as scholars increasingly identified the basic character of contemporary savage mythology with Greco-Roman mythology; for it seemed clear that the myths of Africa, the New World, and Oceana were not intended to carry deeply hidden moral or spiritual meaning. They were much more frequently imaginative stories told about the origins of aspects of the natural world or of human traditions.

In place of the allegorizing approach, Bayle introduced a twofold interpretation of ancient myth and religion. The first and least problematic approach came from the Greek author Euhemerus, who had argued that many of the ancient gods and stories about the gods were grounded in the noteworthy accomplishments of ordinary human beings. These stories were exaggerated and embellished either to solidify the power of the person who was the central character, or to make the story easier to remember. Over time, the original persons and events commemorated might be lost to memory, leaving nothing but a mythic residue. Most eighteenth-century scholars who accepted Euhemerist interpretations of some mythology also accepted other approaches, but they argued that many myths could be connected to historical events (Manuel 1959, 106–107).

In addition to the Euhemerist tradition, Bayle promoted a more complex and ultimately more important anthropological tradition that had its roots in the ancient materialist doctrines of Democritus, Epicurus, and Lucretius. According to this tradition, early religions were associated with human fear and awe in the face of powerful and often unusual natural phenomena. The gods were thus projected as extremely powerful, that they might produce such marvels. From its beginnings, this tradition insisted that what was projected as the consequences of divine actions actually arose out of ignorance of the true causes of events. Epicurus therefore suggested that the most effective way to avoid the fears associated with belief in the gods is to learn the natural causes of phenomena. To all of these notions he added a sense that the priests of ancient religions exploited the fears and gullibility of the people in order to line their own pockets and perpetuate their own power.

To the extent that the dominant classical interpretation of the origins of religion and religious practices was critical of religion in general, of superstition, and of priestly authority within pagan religions, those criticisms indirectly encouraged skepticism and anti-clerical sentiment among Christians. A few seventeenth-century figures openly extended classical pagan attacks on priestly elites to Christian practices, as when Thomas Hobbes criticized the beliefs and practices associated with the Catholic ritual of exorcism in *Leviathan* (1651). Speaking of Catholic priests, he writes:

Who that is in fear of ghosts, will not bear great respect to those who can make the holy water that drives them from him. . . . By their demonology, and the use of exorcism, and other things appertaining thereto, they keep, or think they keep, the people in awe of their power . . . [and] whatsoever ecclesiastics take upon themselves, . . . in their own right, though they may call it God's right, it is but usurpation." (Shapin and Shaffer 1985, 96–97)

Even before Hobbes criticized demonology as a strategy used by Catholic priests to enhance their power, however, there had been a fascinating practical application of this notion connected with French king Henry IV's 1588 Edict of Nantes, which granted Protestants freedom in certain French towns. The goal of Henry's policy was to stamp out conflict between Catholics and Protestants by encouraging mutual toleration. A group of Jesuit priests caused trouble, however, by going from town to town staging spectacular public exorcisms of demons from a woman named Marthe Brosier, in an attempt to demonstrate the power of Catholic ritual and thereby convert Protestants into Catholics.

Henry directed a panel of medical experts to investigate the exorcisms in order to determine whether they were fraudulent or whether Marthe was simply mentally ill and being exploited by the priests. The resulting *Discours véritable sur la faict de Marthe Brosier de Romorantin prétendue démoniaque* (True Discourse about the Facts in the Case of Martha Brossier, who Claimed to be Posessed by Demons), which appeared in 1599, established a series of tests to determine whether someone was possessed or merely mentally ill. It subjected Marthe to these tests, and it concluded that she was a deluded and psychologically imbalanced woman, who had been exploited by her family for financial gain and by the priests for the seditious purpose of fomenting anti-Huguenot feeling. It summed up the exorcisms as performances that involved "nothing from the devil, much counterfeit, a little from disease" and suggested that many persons believed to be possessed, including most of those accused of witchcraft, were really just mentally ill (Walker 1981, 35). The work was immediately translated into English, and it seems to have played a role in slowing down persecutions for witchcraft at the beginning of the seventeenth century in both France and England. Moreover, it proved to be the first of a rapidly growing tradition of scholarly works that treated religious beliefs as superstitious psychopathologies, which culminated in the publication of John Trenchard's *Natural History of Superstition* in 1709.

Departing slightly from the classical anthropological tradition and the medical/psychological tradition, both of which tended to see human nature as constant over time, was a tradition that claimed that there was a progress over time in the human mind, from a primitive mentality, incapable of abstract thought and limited to thinking in terms of parables and myths, to the modern rational mentality, capable of abstraction and suited to stating principles in the form of general propositions. In *The Divine Legation of Moses* (1738–1741), Bishop William Warburton used the notion that primitive tribesmen were unable to think abstractly to account for the way in which the Bible anthropomorphized God (Manuel 1959, 142).

Auguste Comte brought together the ancient atomist understanding of religion with the progressive theories that had emerged in the eighteenth century in what he called the Law of Three Stages:

> Each of our leading conceptions—each branch of our knowledge—passes successively through three different theoretical conditions: the Theological, or fictitious; the metaphysical, or abstract, and the Scientific, or positive. In other words, the human mind, by its nature, employs in its progress three methods of philosophizing, the character of which is essentially different, and even radically opposed.: viz., the theological method, the metaphysical, and the positive. (1974, 25)

Since for Comte, theology and science represent two different and incompatible methods for achieving the same ends—that is, understanding natural events—it was, of course, inevitable that they should be in competition and conflict with one another. Not everyone, however, was or is willing to grant Comte's initial premise, drawn from the ancient Epicurean tradition, that the aims of science and religion are identical.

RELIGION AND THE EMOTIONS

Though David Hume concurred with much of the ancient atomist-inspired anthropological treatment of the origins of religion, he began to turn away from the idea that religion was initiated in connection with natural phenomena. Instead, he focused attention on the way in which religious rituals tend to emphasize events important in the lives of humans:

the first ideas of religion arose not from a contemplation of the works of nature, but from a concern with regard to the events of life, and from the incessant hopes and fears, which actuate the human mind. Accordingly we find, that all idolaters, having separated the provinces of their deities, have recourse to that invisible agent, to whose authority they are immediately subjected, and whose province it is to superintend that course of actions, in which they are engaged. JUNO is invoked at marriages; LUCINA at births. NEPTUNE receives the prayers of seamen; and MARS of warriors. . . . Agitated by hopes and fears . . . men scrutinize, with trembling curiosity, the course of future causes, and examine the various and contrary events of human life. (1993, 139–140)

Hume agreed with earlier anthropologists that primitive men tended to attribute human characteristics, including wants, needs, love, and malice, to natural objects, and that they came to "acknowledge a dependence on invisible powers, possessed of sentiment and intelligence" (Hume 1993, 142). For Hume, however, the emphasis is always on the relationship of the divine entity to some human dependency, rather than on an understanding of the object as a part of the natural order. Finally, Hume insisted that any human emotion, hope as well as fear, and gratitude as well as disappointment, may stimulate religious sentiments, though he did agree with the ancients that fear and terror seem the strongest emotions.

Among the few thinkers who took seriously David Hume's claims that religion was primarily aimed at meeting a range of emotional needs associated with human activities was a young German who underwent a religious conversion while traveling on a diplomatic mission in England in 1758. Turning his back on the rationalism of most Enlightenment figures, Georg Hamann (1730–1788) argued that the path from a primitive mentality to enlightened rationalism represented a degeneration rather than progress. The concrete image is more powerful, creative, and capable of arousing the passions than an abstract representation, even for persons living today, he insisted. This is the reason that God revealed himself through parables and vividly concrete images rather than through bland prose: "All mortal creatures are able to recognize the truth and essence of things only in parables" (Manuel 1959, 287).

Hamann's views were spread widely among German intellectuals by his student, Johann Gottfried Herder. Like his mentor, Herder argued that there are important ways in which early poetic language, before it becomes written and abstract, is more powerful and expressive than written, prosaic language:

> As long as language is not yet a book language, but the language of song, it has a wealth of images and the most exalted harmony. As it becomes the language of civilized people, it gains a greater wealth of political expression, but the exalted harmony and the fullness of the images are toned down. As a *book language* it becomes richer in concepts, but the poetic harmony turns into prose; the image becomes parable, the vivid *ringing* words disappear. As a *philosophical* language it becomes precise, but impoverished. . . . *Poetically,* a language is most consummate *before* it is written; philosophically, when it is written *only;* it is most *useful* and *convenient* when it is both spoken and written. (Zammito 2002, 159, emphasis Herder's)

Herder's interest in the early linguistic productions of a culture was intensified because he believed that each culture was unique as a consequence of the specific environment that produced it. Furthermore, he believed that this uniqueness was most profoundly expressed in a culture's earliest linguistic productions—the folk songs and local mythologies that were still preserved in the countryside where the homogenizing pressures of civilization had not yet penetrated. Thus, Herder began to collect and publish German folk songs, and he encouraged others, including the Grimm brothers, Jakob and Wilhelm, to collect the folk tales that virtually everyone now knows as Grimms' fairy tales.

Herder's insistence that the early poetic literature of a people expresses most completely the unique character of that people led him to approach the Bible in a new way, which has been particularly clearly described by Frank Manuel:

> Herder's elucidation of this "oldest document of the human species" was unlike any Scriptural commentary which had ever been composed before. It was not a reconciliation of contradictions in the manner of the rabbis; nor was it an interpretation of the cosmography of Genesis in the light of astrophysics; nor did it borrow from the textual analysis of the new-born seventeenth and eighteenth century higher criticism. Herder studied the Bible for its graphic portrayal of the historic primitive Hebrews. (1959, 286)

Herder's method of interpreting myth and Scripture—what modern scholars call the hermeneutic method—was historical in a double sense; for Herder emphasized not only the fact that one must approach a text by understanding as best one can what its generators intended in the particular context in which it was created, but also the fact that interpreters bring to their tasks the values and interests of their own cultures. In particular, Herder was concerned that modern interpreters of Scripture recognize and, to the extent possible, rise above, their preferences for scientific or philosophical language. Newton had failed to do this, writing that poetry was the "infant rattle of primitive mankind" (Stanford 1980, 45). For him it was a mode of communication incapable of fully expressing the fundamental philosophical truths that underlay all primitive, and therefore, imperfect, attempts to articulate them. Herder, and subsequently theologians such as Schleiermacher, argued that ancient mythology and the stories contained in the Bible expressed perfectly what they were intended to; this just did not happen to be the truths articulated by modern prose philosophy and science.

It was possible that the stories of the Bible and some modern philosophical treatise might deal with the same topic—for example, the relationship between humans and the divine. But the Bible did so in the form of moving parables and illustrative stories, while philosophy, including theology, did so directly in terms of prose propositions. Poetic, mythic expressions were not failed philosophy, to be rejected as science and philosophy conquered ever greater domains of experience, as Comte later insisted. They were instead a completely different mode of expression with their own enduring advantage over the philosophical.

David Sloan Wilson has offered a modern evolutionary explanation of why a religion built upon the ideas of people such as Hamann and Herder should persist and flourish even within a society in which reason and scientific thought have become dominant for most purposes. He writes:

> Even massively fictitious beliefs can be adaptive, so long as they motivate behaviors that are adaptive in the real world. At best, our vaunted ability to *know* is just one tool in a mental tool kit that is frequently passed over in favor of other tools. . . . From this perspective, we should expect moral systems to frequently depart from narrow reasoning on the basis of factual evidence. Once this kind of reasoning is removed from its pedestal as the only adaptive way to think, a host of alternatives become available. Emotions are evolved mechanisms for motivating behavior that are far more ancient than the cognitive processes typically associated with scientific thought. . . . We might expect stories, music, and rituals to be at least as important as logical

arguments in orchestrating the behavior of groups. Supernatural agents and events that never happened can provide blueprints for action that far surpass factual accounts of the natural world in clarity and motivating power. These otherworldly elements of religion cannot completely eclipse scientific modes of thought, which are superior in some contexts, but the reverse statement is equally true. (2002, 41–42)

From this perspective, though they certainly did not understand what they urged in the same way that a modern secular evolutionary theorist might, Herder and Hamann promoted forms of expression—including religious expression—that call forth deeply ingrained human responses, which logical arguments and abstract principles are simply incapable of touching. No matter how many times one might say, for example, "treat all persons as your neighbors," that statement cannot have the same emotional impact as telling the story of the Good Samaritan.

Subsequent theorists—especially followers of Schleiermacher—would say that the Good Samaritan story is a "Vorstellung," a representation of the "Begriff," or concept, expressed in the statement that one should treat all persons as neighbors. Moreover, they would argue that the representation virtually always has a more powerful emotional and motivational effect. Philosophy, which deals in concepts, may thus consider the same subject matter as religion, which deals in representations. But one is not a substitute for the other, and religion will always be more important for the vast majority of human beings because of its ability to tap into our emotions.

IMMANUEL KANT'S SEPARATION OF SCIENTIFIC KNOWLEDGE FROM RELIGIOUS FAITH

Herder and Hamann's assertion that poetic expressions and myths—including much of the text of the Bible—had different aims from modern science and academic philosophy found a philosophical justification in the "critical philosophy" of Immanuel Kant (1724–1804). Born the fourth of nine children of a Pietist harness maker in East Prussia, Kant studied at Königsberg with Martin Knudtzen, one of the first German professors to teach Newtonian natural philosophy, and developed an initial reputation as a natural philosopher. He wrote a paper on *vis viva* (what we now call kinetic energy) for the Prussian Academy of Sciences in 1747. Then, in 1755, he produced the first major theory of cosmological evolution, *A General Natural History and Theory of the Heavens, or an Attempt to Explain the Composition and Mechanical Origin of the Universe on Newtonian Principles*. But Kant's interests were amazingly wide-ranging, encompassing aesthetics, religion, anthropology, and physical geography, as well as natural history, mathematics, and epistemology (the theory of how we know what we know); between 1755 and 1781, he taught a wide range of courses and became a famous and popular teacher.

Then, between 1781 and 1793, he published a series of works that, taken together, comprise the "critical philosophy." These works attempted to analyze the foundations of virtually all domains of human thought and action. He began with the *Critique of Pure Reason* (1781), which dealt at a very abstract level with how we organize our experience. He then moved on to *A Prolegomena to Any Future Metaphysics* (1783), which sought to illustrate the principles of pure reason by applying them to the foundations of mathematical knowledge; *The Metaphysical Foundations of Natural Philosophy* and *Fundamentals of the Metaphysics of Morals* (1785), which dealt with the underpinnings of natural science and morality respectively; a *Critique of Practical Reason* (1788), which continued the discussion of ethics and morality; the *Critique of Judgment* (1790), which focused on aesthetics; and then to *Religion within the Limits of Reason Alone* (1793), which completed what many subsequent thinkers have characterized as Kant's Copernican revolution in philosophy.

Even Kant admitted that his philosophy was difficult to understand; but it is important to be familiar with a few of his most fundamental ideas in order to follow many nineteenth-century arguments about the relationships between science and religion. First, according to Kant, pure, or "speculative," reason is capable of producing certain knowledge only about "phenomena"—that is, things as they are given to us by our senses and as those sensations are organized by our faculty of understanding or intellect. This faculty structures our experiences according to a set of categories, among the most important of which are space, time, substance, and causation.

The world of our experiences, which is the world of the natural sciences is thus spatial, temporal, filled with substances, and causal, not because this is the case of the world underlying those experiences—what Kant called the "noumenal" world, or world of "things as they are in themselves (*ding an sich*)"—but rather because our mental apparatus forces us to think of them in that way.

Since this idea is likely to seem very strange, consider the following analogy. Both humans and frogs see as a consequence of the fact that photons fall on light sensitive receptors at the back of the eye; but the visual system of the frog processes the incoming light in a different way than the human visual system does. The human system sends a signal to the brain only after several photons have fallen on the same small receptor, whereas the frog system sends a signal to the brain only when photons fall on a sequence of spatially distinct receptors. As a consequence, while humans are able to "see" objects that do not move in their visual field, frogs only see objects that move; so the frog's phenomenal world contains no stationary objects, even though a human, looking at the same part of the universe, may see many. Because the frog subsists by catching flies and other insects that move, this system works well for the frog by focusing attention on its potential prey and screening out irrelevant and possibly distracting information. The important thing to note, however, is that the world of the frog's experience is different from that of human experience because the mental apparatus of frogs is different from that of humans.

Given the fact that our mental apparatus structures experience and does not simply mirror what exists independent of us, Kant argued that universal natural laws could, in some important sense, be understood as following from the laws of our understanding. He concluded the *Prolegomena* with a very famous statement: "It thus at first sounds strange, but it is none the less certainly true, if with regard to [universal natural laws] I say: *The understanding does not extract its laws from, but prescribes them to nature*" (1902, section 36). One of the most important consequences of this notion is that because the mind forces us to think causally about phenomena, the world of phenomena and of science must be deterministic—nothing can happen that is not the consequence of a specific cause. Thus, the world of science does not and cannot in principle, have room for truly free choice.

On the other hand, for Kant our personal experience of freedom is so immediate and powerful that it simply cannot be denied. Furthermore, it is tightly bound up with the very possibility of moral responsibility and thus, of religion. If that is true, however, the domain of morality and religion cannot be the domain of phenomena and science, since there can be no choice in the scientific world. Science, which is the certain creation of speculative reason can quite literally have nothing to offer religion, which depends on something that the world of science cannot contain—freedom. Thus, Kant writes:

I maintain that all attempts to employ reason in theology in any merely speculative manner are altogether fruitless and by their nature null and void, and that the principles of its employment in the study of nature do not lead to any theology whatsoever. . . . [With respect to religion] I have therefore found it necessary to deny knowledge in order to make room for faith. (Kant 1965, 528–529)

Though he approached the issue from a very different perspective than Hamann and his student, Herder, Kant thus ended up supporting their insistence that religious discourse was about a different realm of experience than scientific discourse, denying in the process that natural theology could have any validity.

THE POST-KANTIAN TRADITION IN GERMAN THEOLOGY—SCHLEIERMACHER AND HEGEL

Two major and related issues growing out of Kantian philosophy had a huge impact on early nineteenth-century science and religion. First was Kant's suggestion that natural laws were somehow imposed upon nature, rather than drawn from it. Though Kant clearly meant this notion in a fairly restrictive sense, others were inclined to extend it to grant a much greater role to speculation and imagination—even in the sciences and philosophy—than they had previously held. The distinguished Irish mathematician and physicist William Rowan Hamilton expressed this more liberating view, writing that physical scientists

aim to assign links between reason and experience; not merely by comparing some phenomena with others, but by showing an analogy to those phenomena in our own laws and forms of thought, "darting our being through earth, sea, and air. . . ." And this appears to me to be an essentially imaginative process; although I do not deny that it must be combined with a diligent attention to the appearances themselves in their minute details, and with rigorous reasoning on the hypotheses the scientific imagination has suggested. (Hankins 1980, 104)

Hamilton's statement, like Kant's initial discussion, makes a distinction between the phenomenal world and our accounts or understandings of that world. In addition, it suggests a greater role for the human imagination in constructing scientific hypotheses, which must then be shown to be consistent with experience. Many scientists, especially on the Continent, took these ideas very seriously and granted much greater speculative freedom to scientists than previous notions of science would have allowed.

In *The Critique of Judgment,* Kant posited the existence of another mental faculty, the faculty of judgment, which, unlike the faculty of the intellect, was not necessary to make experiences possible. It was, however, necessary to integrate our particular experiences into a single coherent whole. In Kantian terms, this faculty has a regulative, rather than a constitutive function relative to experience. We could, in theory, have a set of experiences that were totally unrelated to one another; but such a set of experiences would be both unsatisfying and useless—incapable of producing what we call knowledge, which can alone be used to guide future actions. Our faculty of judgment, by insisting upon a notion of purposiveness, allows us to integrate our experiences by viewing them as directed by a single plan. According to Kant, but not most of his followers and critics, we cannot know that there is purpose outside of ourselves; we must simply formulate our knowledge as if there were. Even more than his earlier notions, the idea that a faculty of judgment regulates our knowledge seemed to warrant the claim that all knowledge is inescapably subjective.

As applied to the general problem of unifying the deterministic world of phenomena with the non-deterministic world of noumena, Kant's *Critique of Judgment* was widely held to be a failure, and attempts at remedies led in several different directions. The most important of these for an understanding of science and religion relationships was developed by F.W.J. Schelling, who had been trained as a theologian before turning his attention to medicine, physics, and mathematics. Schelling argued that there must be something more fundamental than either the subjective world of individual inner experience or the objective world of collectively experienced material existence. This more fundamental something he called the "Absolute" in the 1803 edition of his *Ideas for a Philosophy of Nature,* and he argued that this Absolute manifests itself in two different ways: as Mind in us, and as Nature outside of us. The fact that Nature and Mind are both manifestations of the same Absolute accounts for why our mental faculties are matched to the structure of the natural world.

Many German theologians and philosophers, including Schleiermacher and G.W.F. Hegel, the most highly regarded German philosopher of the early nineteenth century, appropriated Schelling's basic ideas and identified his Absolute with the Christian God. But Schleiermacher and Hegel emphasized very different aspects of Schelling's work. Schleiermacher followed Schelling in arguing that we recognize the same kind of complete dependence on the Absolute as spiritual beings that we recognize with regard to the natural world as physical beings. It is this feeling of complete dependence that grounds all religion. Thus, religion is born out of pre-cognitive and even pre-volitional feelings according to Schleiermacher. It is these feelings that the poetry and stories of the Bible are capable of reaching. Though Schleiermacher continued to be relatively orthodox in his Christian beliefs, identifying Jesus of Nazareth with the savior Christ, he did argue that the miracles reported in the Bible as evidence of scriptural authenticity were simply irrelevant to Christian belief. God's presence could not be "proven" at all, it simply had to be felt in ordinary events. The effect of Schleiermacher's views was thus to reinforce the separation of religion from the domains of reason and science, which had been promoted by Hamann, Herder, and Kant.

Hegel, on the other hand, focused on the accessibility of God by human reason as a consequence of the fact that humans participate in the Absolute, or Idea, by virtue of their intellect. He violently objected to Schleiermacher's emphasis on the feelings of dependency as a foundation for religion, on the grounds that if Schleiermacher were right, "a dog would be the best Christian because it has this feeling [of dependence] most intensely" (Gregory 1992, 37). A much more rigorous thinker than Schleiermacher, Hegel gave rise to a tradition of speculative theology that viewed the history of the natural and human worlds as a process by which the Idea becomes increasingly self-conscious through its interaction with nature.

On the other hand, Hegel did admit that myth and Scripture addressed themselves to a different side of humanity than rational, or speculative, theology. It was Hegel, for instance, that made explicit the notion that, although religion and philosophy might have the same subject matter, religion had to express its content in the form of symbols or representations, while philosophy expressed its content directly through concepts.

A NEW ANTHROPOLOGY OF RELIGION—FEUERBACH

A theology student of Schleiermacher's who was fascinated by Hegelian philosophy pushed the distinction between religious representation and conceptual expression in a new direction that would have been condemned by both of these philosophers. This young theologian-philosopher, Ludwig Feuerbach (1804–1872), re-integrated scientific knowledge and religion by establishing an anthropological approach to religion that drew from—but radically transformed—the classical anthropological tradition.

In order to understand what Feuerbach attempted and why he attempted it, we need briefly to discuss the relationship that had developed between Lutheran evangelical Christianity and the German states in the early nineteenth century. From near its inception, Lutheran religion and the German states had a very close relationship. At least in part to garner political support, Luther had emphasized the Christian's obligation to obey secular authority regarding worldly things, insisting on one of the few statements by Jesus that is reported identically in the Gospels of Matthew (20:21), Mark (12:17), and Luke (20:25)—that is, "Render unto Caesar that which is Caesar's and render unto God that which is God's." From this perspective, religion was primarily about salvation in an afterlife, while the church accepted a secular role as a promoter of political obedience and conformity.

After the German revolts against Napoleonic occupation in 1813, the symbiotic relationship between church and state was formalized when the clergy were made paid state employees by most of the thirty-nine signatories of the 1815 pact establishing the German Confederation, which included the vast majority of the German population, and whose leadership was socially conservative. At the same time, a nationalist movement that was especially popular among university students and faculty not only threatened the traditional aristocracy by advocating the creation of a nation that would undermine the political autonomy of the various local states; it also promoted a democratic mood by suggesting that the aristocracy—whose culture was strongly influenced by French literature and fashion—was somehow inferior to ordinary Germans—Herder's *Volk*. Under these circumstances, state governments, often aided by local religious figures, tried to repress the nationalist movement, suppressing student organizations and public celebrations of German culture and jailing those who dissented. In addition, neither the governments nor the churches seemed particularly sympathetic to the growing economic distress felt by German workers and artisans in the second and third decades of the nineteenth century.

Those who felt strongly that Christianity should promote human welfare in this world—a message that had dominated Christian humanist writings—were deeply disturbed by what seemed to be a kind of callousness regarding human distress that had overtaken the traditional evangelical church. Feuerbach began to give these feelings a voice in 1829 when he wrote a short pamphlet, *Thoughts on Death and Immortality*, making the distinction between the symbolic or representational character of religious expression and the direct, conceptual character of philosophical discourse. According to Feuerbach, stories of heaven and immortality were representations of our search for perfection in life. When they were taken literally rather than symbolically, as the evangelical Lutherans did, instead of serving to inspire Christians to accept their obligations to help one another in this life, they encouraged people to turn away from their obligations and to hold off their hopes for perfection for the afterlife. So, in *Thoughts on Death and Immortality* Feuerbach sought "to cancel above all the old cleavage between this side and the beyond in order that humanity might concentrate on itself, its world, and its present with all of its heart and soul" (Gregory 1977, 16).

Soon after he wrote *Thoughts on Death and Immortality*, Feuerbach began to focus on concrete physical conditions as the source of our mental life, rather than on our ideas as productive of the material world. Feuerbach revealed his new philosophical perspective in 1841 in *The Essence of Christianity*, which shocked and rocked both the religious and the philosophical communities; for in it Feuerbach proposed a special empirical, anthropological approach to understanding religion. In the preface, he outlines his new method—which repudiates Kant and the Idealist tradition that Kant's work stimulated—both clearly and passionately:

> I unconditionally repudiate *absolute*, immaterial, self-sufficing speculation,—that speculation which draws its materials from within. . . . For my thought, I require the sense, especially sight; I found my ideas on materials which can be appropriated only through the activity of the senses. I do not generate the object from the thought, but the thought from the object; and I hold that alone to be an object which has an existence beyond one's own brain. . . . I attach myself, . . . only to *realism*, to materialism in the sense above indicated. . . . I am nothing but a *natural philosopher in the domain of the mind*; and the natural philosopher can do nothing without instruments, without material means. In this character I have written the present work, which consequently contains nothing else than the principle of a new philosophy verified practically . . . in application to a special object, but an object which has a universal significance: namely religion. (1957, xxxiv, emphasis Feuerbach's)

In this work, and in *The Essence of Religion*, which followed in 1851, Feuerbach continued the project that he had begun with *Thoughts on Death and Immortality*. He sought to make "the friends of God into the friends of man, believers into thinkers, worshipers into workers, candidates for the other world, into students of this world, Christians, who on their own confession are half-animal and half-angel, into men—whole men, . . . [and] religious and political footmen of a celestial and terrestrial monarchy and aristocracy into free, self-reliant citizens of earth" (1957, xi). In order to accomplish this goal, Feuerbach had to turn the traditional Christian notion of God on its head and argue that God did not create us but that, instead, we created God. Like those in the classical anthropological tradition, Feuerbach argued that we create God by projecting human characteristics into the universe. But unlike the older tradition, he did not argue that this projection is a response to fear of natural phenomena. Instead, for Feuerbach, we create God by projecting our ideal selves—selves that we are too imperfect to attain—into some transcendent domain so that we might have some image in mind as we strive to make ourselves better humans.

Once again, as in his earlier work, Feuerbach insisted that, so long as we recognize what we are doing and use that ideal self as a means for motivating Christlike behavior toward our fellow humans, this human creation serves positive ends. But when we forget and grant independent and transcendent existence to God, as we inevitably do, making him a lord over and above humanity, as in traditional evangelical Christianity, we diminish ourselves. Thus, he writes:

> [the important question is] not whether God is a creature whose nature is the same as ours but whether we human beings are to be equal among ourselves; not whether and how we can partake of the body of the Lord by eating bread, but whether we have enough bread for our own bodies; . . . Not whether we are Christians or heathens, theists or atheists, but whether we are or can become men, healthy in soul and body, free, active, and full of vitality. . . . I deny God. But that means for me that I deny the negation of man. In place of the illusory, fantastic, heavenly position of God, which in actual life necessarily leads to the degradation of man, I substitute the tangible, actual, and consequently also the political and social position of mankind. (1969, 17)

Feuerbach's views were clearly antithetical to traditional evangelical Christianity, and they were especially troublesome to social and political conservatives, who viewed the promise of rewards and punishments in an afterlife as a powerful tool for enforcing obedience to secular authority; but they were widely appealing to a younger generation of politically liberal Germans who chafed under repressive political regimes throughout the Germanies.

DAVID STRAUSS AND THE USE OF SCIENCE TO REJECT EVANGELICAL CHRISTIANITY

Among those drawn to Feuerbach's views was another young student of Schleiermacher and leftist Hegelian, David Strauss, who had begun to move in a very similar direction with the publication of his *Life of Jesus* in 1835. Strauss had also become disenchanted with traditional Christianity before he found science and applied it to his critical tasks; and like Feuerbach, he began by focusing on the representational modes of expression appropriate to religion. From this perspective, Strauss argued that, though Jesus of Nazareth was probably a historical figure, few if any of the stories attached to his life in the Bible were likely to be factually true, and none of the miracle stories were credible, taken literally. This did not mean, for Strauss, that these stories could not represent true aspects of the relationships between the human and the Divine. They were fictions or myths, created in a particular context to symbolize one stage in the unfolding relationship of God to man, which Hegel's system posited.

Rather than investigating religion from an anthropological perspective, however, Strauss used his scientific knowledge of geology and biology to directly challenge the literalist interpretation of Scripture favored by many evangelical Christians, beginning in his *Doctrine of Faith*, published in 1840. Commenting on Genesis, for example, he wrote:

> According to science, . . . there is no doubt that our planet has acquired its present state gradually, that it was uninhabitable for organic beings in primitive times, and that these have originated gradually without having had ancestors—that is, through dissimilar reproduction. (Gregory 1992, 80)

In his early works, Strauss continued to think of himself as a Christian theologian, though he lost his appointment in theology at Tubingen and was generally condemned by the traditional religious community. He became embittered, however, and as time went on, he argued that since the process by which the Spirit expressed itself in human consciousness called forth new forms of expression, the old biblical myths no longer represented the most advanced religious views. He turned, instead, to natural science, especially to evolution, to find his religious truths.

AUGUSTE COMTE'S "RELIGION OF HUMANITY"

The 1820s through the 1840s saw the rise of a series of self-styled religions that rebelled against traditional Christianity and turned to science to provide some new guidance for human action. These included the *Religion de Saint Simon,* established in 1829 by two graduates of the *École Polytechnique* and followers of the famous utopian socialist Henri Saint-Simon; the Communitarian Society of Rational Religionists, established in 1835 by Robert Owen, another socialist leader; and the National Secular Society, established by George Holyoake in 1845. Though led by middle-class intellectuals, all of these science-based religions were aimed at working-class audiences and were grounded in a sense that traditional religious institutions were colluding to oppress the poor and powerless. The secularist preacher Robert Cooper probably spoke for them all when he preached in 1853 that

> Willingly or unwillingly on their part [the Christian clergy] are the stumbling block in the way of every effort to enlighten and emancipate mankind. Talk of social reform and they exclaim that poverty is a *divine* ordinance; that God made both poor and rich, and that the people must, therefore, "be content in the situation in which Divine providence has placed them." Talk of *political* reform, and they remind you that it is your duty, by command of the inspired word of heaven, to submit, "to the powers that be." (Royle 1971, 93)

All of these groups hearkened back to the Christian humanist ideas of Francis Bacon and viewed science as the foundation of a more socially responsible approach to the world—one that sought to improve the lives of people here and now, rather than fixing their sights on some distant future afterlife. Some of them drew substantial numbers of people to their services. Secularist preachers, for example, probably preached to a total of 200,000 persons; but secularism as a significant force, was short-lived, as were the Saint-Simonian religion and the Owenite Rational Religionist movement. There was, however one mid-nineteenth-century science-based religion that had a significant long-term impact, the Religion of Humanity established by Auguste

Comte. At roughly the same time that Feuerbach was developing his materialist and anthropological approach to religion, the French mathematics student-turned-philosopher and historian of science Auguste Comte was working his way toward a scientifically grounded religion, which he identified as the "Religion of Humanity."

Comte's first great work, *The Positive Philosophy*, had been written in the 1820s. In it, Comte announced his law of three stages and offered an extended analysis of the historical development and philosophical character of the positive sciences to date, projecting the existence of a not-yet-established social physics or "sociology," which would be capable of placing society on a scientific foundation and bringing order as well as progress to the chaotic social world of the early nineteenth century. By 1851, Comte felt that sociology had advanced sufficiently that the time had come to establish the new social world, and the first step would be to replace current religions and their frequently destructive consequences with a new kind of religion, the characteristics of which he outlined in *The System of Positive Polity: or a Sociological Treatise Instituting the Religion of Humanity* (see Figure 6.1). The reform of religion had to come first because, according to Comte, moral regeneration had to precede social regeneration, and it was religious institutions that set the moral tone for society.

In *The System of Positive Polity*, Comte took a very different direction from that of *The Positive Philosophy*—one that paralleled the emphasis by Schleiermacher on emotion rather than intellect and that of Feuerbach on the role of humans in creating their own divinities and in projecting them onto the external world. Thus, Comte introduced his second major work by arguing that the time had come to favor "the ascendency of the heart over the head," and by writing:

> In the [Positive Philosophy], where the process of scientific preparation is carried to its fullest limit, I have carefully kept the objective method in the ascendant, as was necessary where the course of thought was always preceding from the world in the direction of man. But the fulfillment of this preliminary task, by the fact of placing me in the true universal point of view, involves henceforth the prevalence of the subjective method as the only source of complete systematization, the procedure now being from man outward toward the world. (Lenzer 1975, 311)

Comte argued that the fundamental goal of the new religion must be to strengthen the social passions relative to the egoistic ones by promoting the universal love of humanity. He was certain that the fun-

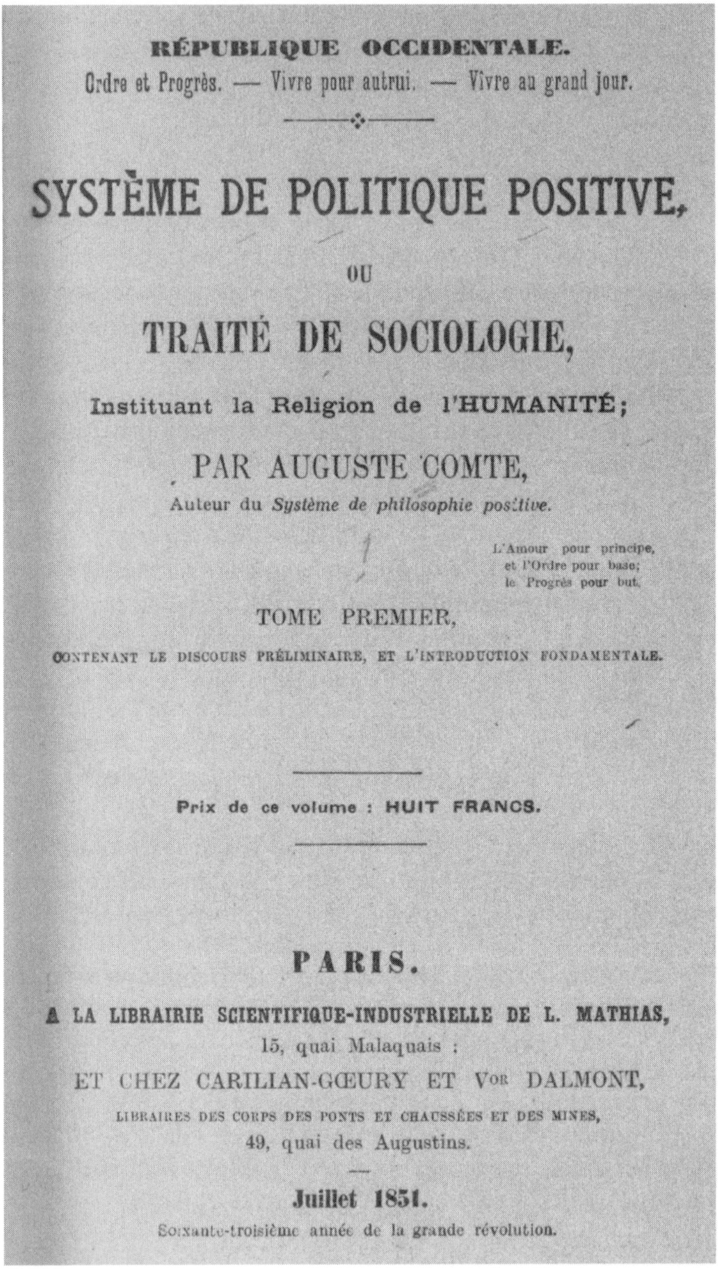

RÉPUBLIQUE OCCIDENTALE.

Ordre et Progrès. — Vivre pour autrui. — Vivre au grand jour.

SYSTÈME DE POLITIQUE POSITIVE,

OU

TRAITÉ DE SOCIOLOGIE,

Instituant la Religion de l'HUMANITÉ;

PAR AUGUSTE COMTE,

Auteur du *Système de philosophie positive.*

L'Amour pour principe,
et l'Ordre pour base;
le Progrès pour but.

TOME PREMIER,

CONTENANT LE DISCOURS PRÉLIMINAIRE, ET L'INTRODUCTION FONDAMENTALE.

Prix de ce volume : HUIT FRANCS.

PARIS.

A LA LIBRAIRIE SCIENTIFIQUE-INDUSTRIELLE DE L. MATHIAS,

15, quai Malaquais :

ET CHEZ CARILIAN-GŒURY ET V[OR] DALMONT,

LIBRAIRES DES CORPS DES PONTS ET CHAUSSÉES ET DES MINES,

49, quai des Augustins.

—

Juillet 1851.

Soixante-troisième année de la grande révolution.

Figure 6.1. Title page of Comte's *System of Positive Polity* in which the "Religion of Humanity" is proposed.

damental tensions between egoism and altruism could never be completely eliminated, because the consequence of pure selflessness would be death. But he did believe that a society built on a healthier balance between the two would be vastly superior to that of his present day. In order to get humans to subordinate their individual wants to some broader set of social goals, Comte argued that they must be made to feel what Schleiermacher had argued that they feel intuitively—a complete dependence on "some external Power possessed of superiority so irresistible as to leave no sort of uncertainty about it" (Lenzer 1975, 396). Fortunately, Comte claimed, biology establishes that just such a dependence on an external power actually exists—the dependency of every organism on its environment to provide both sustenance and the stimuli that initiate physical and mental activity. In the case of humans, this "Supreme Power" is the entire physical and social universe, whose character is made known to us by the positive sciences. Science thus replaces theology in the Religion of Humanity.

If science satisfies our intellectual needs and brings the objects of veneration out of the realm of superstition and into the domain of the natural, there remains a set of emotional needs that must be met through religious worship and ritual. For Comte, as for Feuerbach, humans create that which they worship. But positivists, instead of projecting their own ideal selves onto a divinity to which they then assign independent reality, pick out other individuals—usually dead—who symbolize admired characteristics. The adherents to the religion of humanity then worship the images of the idealized persons and reflect upon the moral attributes that those persons embodied in a particularly clear way. Commemoration of the lives of the dead thus reinforce commitments to the virtues that they symbolize (see Figure 6.2).

The great difference between the positivists' veneration of such figures as Newton, Laplace, Lavoisier, Mozart, and even Jesus, and the Christians' veneration of God, is that the positivists remain conscious of the fact that the figures they worship live only subjectively in memory, not objectively in the external world. "Our beloved dead," Comte wrote, "are no longer governed by the rigorous laws of the inorganic order, nor even of the vital. . . . The existence which each one of them retains in our brains . . . is composed essentially of images, which revive at once the *feelings* with which the being snatched from us inspires in us and the thoughts which he [*sic*] occasioned. Our subjective worship is reduced, then, to a species of internal evocation" (1973, 68).

[B']

CALENDRIER POSITIVISTE

POUR UNE ANNÉE QUELCONQUE

OU

TABLEAU CONCRET DE LA PRÉPARATION HUMAINE, destiné surtout à la transition finale de la république occidentale formée, depuis Charlemagne, par la libre connexité des cinq populations avancées, française, italienne, espagnole, britannique et germanique.

		HUITIÈME MOIS **DANTE.** L'ÉPOPÉE MODERNE	NEUVIÈME MOIS. **GUTTEMBERG.** L'INDUSTRIE MODERNE	DIXIÈME MOIS. **SHAKESPEARE.** LE DRAME MODERNE	ONZIÈME MOIS. **DESCARTES.** LA PHILOSOPHIE MODERNE.	DOUZIÈME MOIS. **FREDERIC.** LA POLITIQUE MODERNE.	TREIZIÈME MOIS. **BICHAT.** LA SCIENCE MODERNE.
Lundi	1	Les Troubadours.	Marco Polo … *Chardin.*	Lope de Vega … *Montalvan.*	Albert-le-Grand. *Jean de Salisbury.*	Marie de Molina.	Copernic … *Tycho-Brahé.*
Mardi	2	Bocace … *Chaucer.*	Jacques Cœur … *Gresham.*	Moreto … *Guillen de Castro.*	Roger Bacon … *Raimond Lulle.*	Côme de Médicis l'Ancien.	Kepler … *Halley.*
Mercredi	3	Rabelais.	Gama … *Magellan.*	Rojas … *Guevara.*	Saint-Bonaventure … *Joachim.*	Philippe de Comines. *Guicciardini.*	Huyghens … *Varignon.*
Jeudi	4	Cervantes.	Neper … *Briggs.*	Otway.	Ramus … *Le cardinal de Cusa.*	Isabelle de Castille.	Jacques Bernoulli. *Jean Bernoulli.*
Vendredi	5	La Fontaine … *Robert Burns.*	Lacaille … *Delambre.*	Lessing.	Montaigne … *Érasme.*	Charles-Quint … *Sixte-Quint.*	Bradley … *Roëmer.*
Samedi	6	Foë … *Goldsmith.*	Cook … *Tasman.*	Gœthe.	Campanella … *Morus.*	Henri IV.	Volta … *Sauveur.*
DIMANCHE	7	**ARIOSTE.**	**COLOMB.**	**CALDERON.**	**SAINT-THOMAS-d'Aquin.**	**LOUIS XI.**	**GALILÉE.**
	8	Léonard de Vinci … *Le Titien.*	Benvenuto Cellini.	Tirso.	Hobbes … *Spinosa.*	Coligny … *L'Hôpital.*	Viète … *Harriott.*
	9	Michel-Ange … *Paul Véronèse.*	Amontons … *Wheatstone.*	Vondel.	Pascal … *Giordano Bruno.*	Barneveldt.	Wallis … *Fermat.*
	10	Holbein … *Rembrandt.*	Harrison … *Pierre Leroy.*	Racine.	Locke … *Malebranche.*	Gustave-Adolphe.	Clairaut … *Poinsot.*
	11	Poussin … *Lesueur.*	Dollond … *Graham.*	Voltaire.	Vauvenargues … *Mme de Lambert.*	De Witt.	Euler … *Monge.*
	12	Velasquez … *Murillo.*	Arkwright … *Jacquart.*	Métastase … Alfieri.	Diderot … *Duclos.*	Ruyter.	D'Alembert … *Daniel Bernoulli.*
	13	Téniers … *Rubens.*	Conté.	Schiller.	Cabanis … *Georges Leroy.*	Guillaume III.	Lagrange … *Joseph Fourier.*
	14	**RAPHAEL.**	**VAUCANSON.**	**CORNEILLE.**	**Le Chancelier BACON.**	**GUILLAUME-le-Taciturne.**	**NEWTON.**
	15	Froissart … *Joinville.*	Stévin … *Torricelli.*	Alarcon.	Grotius … *Cujas.*	Ximenès.	Bergmann … *Scheele.*
	16	Camoens … *Spenser.*	Mariotte … *Boyle.*	Mme de Motteville … *Mme Roland.*	Fontenelle … *Maupertuis.*	Sully … *Oxenstiern.*	Priestley … *Davy.*
	17	Les Romancistes espagnols.	Papin … *Worcester.*	Mme de Sévigné … *Lady Montague.*	Vico … *Herder.*	Colbert … *Louis XIV.*	Cavendish.
	18	Chateaubriand.	Black.	Lesage … *Sterne.*	Fréret … *Winckelmann.*	Walpole … *Mazarin.*	Guyton-Morveau … *Geoffroy.*
	19	Walter Scott … *Cooper.*	Jouffroy … *Fulton.*	Mme de Staël … *Miss Edgeworth.*	Montesquieu … *d'Aguesseau.*	D'Aranda … *Pombal.*	Berthollet.
	20	Manzoni.	Dalton … *Thilorier.*	Fielding … *Richardson.*	Buffon … *Oken.*	Turgot … *Campomanes.*	Berzélius … *Ritter.*
	21	**TASSE.**	**WATT.**	**MOLIERE.**	**LEIBNITZ.**	**RICHELIEU.**	**LAVOISIER.**
	22	Pétrarque.	Bernard de Palissy.	Pergolèse … *Palestrina.*	Robertson … *Gibbon.*	Sidney … *Lambert.*	Harvey … *Ch. Bell.*
	23	Thomas A'Kempis. *Louis de Grenade. et Bunyan.*	Guglielmini … *Riquet.*	Sacchini … *Grétry.*	Adam Smith … *Dunoyer.*	Franklin … *Hampden.*	Boërhaave … *Stahl.*
	24	Mme de Lafayette … *Mme de Staël.*	Duhamel (du Monceau). *Bourgelat.*	Gluck … *Lully.*	Kant … *Fichte.*	Washington … *Kosciusko.*	Linné … *Bernard de Jussieu.*
	25	Fénelon … *Saint-François-de-Sales.*	Saussure … *Bouguer.*	Beethoven … *Handel.*	Condorcet … *Ferguson.*	Jefferson … *Madison.*	Haller … *Vicq-d'Azyr.*
	26	Klopstock … *Gessner.*	Coulomb … *Borda.*	Rossini … *Weber.*	Joseph de Maistre … *Bonald.*	Bolivar … *Toussaint-Louverture.*	Lamarck … *Blainville.*
	27	Byron … *Elisa Mercœur.*	Carnot … *Vauban.*	Bellini … *Donizetti.*	Hegel … *Sophie Germain.*	Francia.	Broussais … *Morgagni.*
	28	**MILTON.**	**MONTGOLFIER.**	**MOZART.**	**HUME.**	**CROMWELL.**	**GALL.**

Jour complémentaire … Fête universelle des **MORTS.**

Jour bissextile … Fête générale des **SAINTES FEMMES.**

Figure 6.2. The last six months of Comte's positivist calendar, indicating the persons to be remembered on each day.

Though never very numerous, advocates of Comte's new religion included a substantial number of important intellectuals throughout the world. These included the historian Richard Congreve, the philosopher Henry Lewes, and the novelist George Eliot, in England; the journalist Girish Chunder Ghose and the high court judge Dwarkanath Mitter, in India; and the republican political reformer and one-time Minister of War and Minister of Education Benjamin Constant, in Brazil. Writing to his wife in 1867, Constant expressed very well and compactly the basic feelings of those who became members of this small but significant sect:

> I am, as you know, a follower of [Comte's] doctrines; I accept his principles and belief: the Religion of Humanity is my religion. I believe it with all my heart. . . . It is a new religion, and rational withal, the most philosophical and only one which flows naturally from the laws which govern human nature. (Costa 1964, 87)

SCOTTISH COMMON SENSE PHILOSOPHY CALLS FOR A SCIENTIFIC RELIGION AND A RELIGIOUS SCIENCE

Virtually all of the claims regarding the nature of science, the nature of religion, and the character of their interactions—or noninteractions—discussed so far in this chapter were opposed by a philosophical movement that goes by the name of the Scottish Common Sense philosophy. Initiated as a rebuttal to the ideas of David Hume by a group of intellectuals from Aberdeen, Scotland, including Thomas Reid, in the works of Reid's successors, Dugald Stewart, Sir William Hamilton, and (for some purposes) John Herschel, it also took on opposition to Kantian tendencies and to materialism. It became extremely important in America through the teaching of the philosophers John Witherspoon, Samuel Tyler, and James McCosh, and of the evangelical "Old School Presbyterian" theologian Charles Hodge.

The most basic claim of the Common Sense school was that Hume, Kant, and those whose ideas flowed from them were fundamentally wrong because they were not sufficiently radical in their empiricism. Hume and Kant alike failed to realize that part of our immediate experience of objects and events in the phenomenal world is a certain, intuitive awareness of their real existence as physical objects and occurrences outside of us. As a consequence, both exercised themselves and spun out convoluted solutions to a pseudo-problem—that of the relationship between objects in the world and our ideas about them. This was a problem for Hume and Kant because we can only think using our ideas; so we want to know how ideas arise in response to our sensory input.

For the Common Sense philosopher, on the other hand, we really do think about the objects outside of us because we experience our core ideas about those objects as perfect images of their real characteristics. Reid expressed the situation as he saw it in 1764 in *An Inquiry into the Human Mind on the Principles of Common Sense* as follows: "[W]hen I perceive a tree before me, my faculty of seeing gives me not only a notion or simple apprehension of the tree, but a belief of its existence . . . and this judgement or belief is not got by comparing ideas, it is included in the very nature of the perception" (1895, 1:209). Dugald Stewart was even more direct, though perhaps a little less precise: "[It is] the external objects themselves, and not any . . . images of these objects that the mind perceives" (Bozeman 1977, 10).

Ultimately, we cannot know why and how our minds accurately perceive the objects of our experience. Neither, however, can we seriously doubt that they do. Furthermore, we really do experience two radically different kinds of real existing entities—material ones and spiritual ones—so dualism is a directly experienced feature of the world. Common Sense philosophy thus opposed those, such as Hobbes or Feuerbach, who would build a materialist philosophy on the back of scientific knowledge. Members of the Scottish Common Sense school did accept certain features of Hume's analysis of causation, arguing that we do not and cannot know anything regarding causes other than the constant relationship of precedence and consequence observed in phenomena. That is, we can seek to discover the laws which describe phenomena over time, but we cannot know how any event or object produces the consequences that it does. This set Common Sense philosophy directly at odds with Kantian developments and those of German Idealists such as Schelling, for whom the discovery of causes was the primary goal of all science.

One consequence of admitting that humans are ultimately unable to understand how phenomena are caused was that it also placed Common Sense philosophy in opposition to the use of imagination and speculation in science. Common Sense philosophers insisted that no mere speculation regarding the relationships among objects of our experiences can have the kind of reliability that we want scientific knowledge to have. All scientists can do is formulate general expressions regarding the patterns that phenomena evidence, grounding these generalizations in the particular sensory experiences that we have. Thus, hypotheses and theories are to be entertained only provisionally and with grave suspicion.

Hypotheses and theories are certainly useful in suggesting new experiments and in forming a kind of scaffolding to stand on while building genuine scientific knowledge; but according to most Common Sense adherents, they should not be misinterpreted as constituting scientific knowledge itself. In its most extreme American version, represented in Samuel Taylor's *A Discourse of the Baconian Philosophy* (second edition, 1846), this philosophy revived and amplified the Baconian emphasis on the primacy of facts and on the induction of more general facts, or laws, from particular ones. Moreover, it explicitly decried invention and speculation. Logical deductions could play a legitimate—but subordinate—role in science only when the premises of a deductive argument were well established inductively or when they were unchallengeable "common sentiments" of humankind, such as "the whole is equal to the sum of its parts."

The Baconian, inductivist method was then to be applied both to the study of nature and to the study of the Bible. When James S. Lamar published his *Organon of Scripture: or the Inductive Method of Biblical Interpretation* in 1859, he argued that "the Scriptures admit of being studied and expounded upon the principles of the inductive method; and . . . when thus interpreted they speak to us in a voice as certain and unmistakable as the language of nature heard in the experiments and observations of science" (Noll 2000, 301). In a similar vein, Charles Hodge, who was professor of theology at Princeton, introduced his *Systematic Theology* (1872–73) in the following way:

> The Bible is to the theologian what nature is to the man of science. It is his storehouse of facts; and his method of ascertaining what the Bible teaches is the same as that which the natural philosopher adopts to ascertain what nature teaches. . . . The duty of the Christian Theologian is to ascertain, collect, and combine all the facts which God has revealed concerning himself and our relation to him. These facts are all in the Bible. (Noll 2000, 301)

Common Sense philosophy was not only consistent with a continuing tradition of natural theology, so long as it was facts rather than theories about nature from which inferences regarding design were drawn, but its advocates also strongly promoted scientific investigation as a religious activity. Edward Everett, the Unitarian editor of the *North American Review* and a staunch promoter of the Scottish version of Baconian philosophy, insisted that "[t]he great end of all knowledge is to enlarge and purify the soul, to fill the mind with noble contemplations, to furnish a refined pleasure, and to lead our feeble reason from the works of nature up to its great author" (Bozeman 1977, 78). The American evangelical scientist Joseph LeConte, who was to become president of the American Association for the Advancement of Science and first president of the University of California, expressed the common view particularly well in introducing a lecture at the Smithsonian Institution in 1857: "Nature is a book in which are revealed the divine character and mind. Science is the human interpreter of this divine book, [it consists of] human attempts to understand the thoughts and plans of the deity" (Bozeman 1977, 80).

Many Common Sense-trained scientists emphasized the religious implications of their work and promoted natural theology. Henry Brougham, for example, who was a student of Dugald Stewart, the author of numerous papers on physical optics, and a founding member of the British Association for the Advancement of Science (as well as a powerful politician), brought out a new and extensively annotated five volume edition of William Payley's *Natural Theology* (1835–1839). Edward Hitchcock, a distinguished American geologist and the president of Amherst College, wrote *The Religion of Geology and its Connected Science* in 1855. James Clerk Maxwell, a student of William Hamilton, a member of the evangelical Free Church of Scotland, and perhaps the greatest physicist of the nineteenth century, made a special effort to promote his natural philosophy course at Aberdeen for theology students:

> Those who intend to pursue the study of Theology will also find the benefit of careful and reverent study of the order of Creation. They will learn that though the world we live in, being made by God, displays his power and his goodness even to the careless observer, yet that it conceals far more than it displays, and yields its deepest meanings only to patient thought. (Jones 1973, 81)

And the important Victorian physicists Peter Guthrie Tait and Balfour Stewart published a fascinating work of natural theology, *The Unseen Universe; or Physical Speculations on a Future State* in 1875. In this work, both the law of conservation of energy and the second law of thermodynamics are assumed to be revisible approximations so that small amounts of energy can be allowed to cross over from the material universe to the unseen spiritual world beyond.

Given the close linkages between Common Sense philosophy, Baconian inductivism, and the widespread American and Scottish view of natural science as a religious calling, it should hardly be surprising that when advocates of certain kinds of science promoted views that seemed in violation of both their religious beliefs and the canon of proper inductive methods, American and Scottish believers were highly critical.

In America, scientists and theologians across most of the religious spectrum, from moderately conservative Presbyterians to ultra-liberal Unitarians, seemed to agree that there could not be any ultimate conflict between God's two books—that of nature and that of Scripture; but Presbyterians were inclined to read God's word much more literally than Unitarians, who gradually adopted Schleiermacher's hermeneutical approach. By 1849, for example, the Unitarian Francis Bowen could write in opposition to his Presbyterian challengers, "the literal interpretation of the first chapter of the book of Genesis has come to be regarded by nearly all educated Christians in the same light with the Papal opposition to the doctrine that the earth revolves round the sun" (Roberts 1988, 24). Charles Hodge, the chief Presbyterian theologian of his age, on the other hand, was still promoting a primarily literal reading of Scripture in the 1870s. Thus, Hodge argued that scientists should be "cautious in announcing results . . . even apparently hostile to the generally received sense of Scripture" (Bozeman 1977, 109).

Chapter 7

Back to the Beginnings—of the Earth, of Life, and of Humankind, 1680–1859

Many religious conservatives and evangelicals in England and virtually all major Protestant theologians on the Continent turned away from natural theology during the late eighteenth and the nineteenth centuries, moving either to focus on revealed religion or to explore an anthropological understanding of religious developments that emphasized the value of religion in the here and now. But there was a new—or newly revived and reworked—tradition within natural theology that dominated science and religion interactions in Britain throughout the eighteenth and the first half of the nineteenth century and which was at least present as a minor theme on the Continent.

Hexamera, or commentaries on Genesis, which focused on the creation of the universe, the earth, the various forms of plant and animal life, and human kind, had dominated Christian natural theology from the second through the sixteenth centuries, but the great British natural theologians of the seventeenth century—Charleton, Wilkins, Boyle, Cudworth, and Ray—had abandoned that pattern in favor of a more static investigation of the evidence for design in the present state of the world. Starting with Thomas Burnet's *Sacred Theory of the Earth*, published in five installments between 1680 and 1689, and his *Archaeologiae Philosophicae* of 1692, there was a new emphasis on temporal change. Like the older hexamera, Burnet's works and the roughly thirty new works immediately responding to Burnet's—including John Woodward's *Essays Toward a Natural History of the Earth* (1695) and William Whiston's *A New Theory of the Earth, from Its Orig-*

inal, to the Consummation of all Things, Wherein the Creation of the World in Six days, the Universal Deluge, and the General Conflagration, As laid down in the Holy Scriptures, Are shewn to be perfectly agreeable to Reason and Philosophy (1696) (see Figure 7.1)—used a biblical framework to organize their scientific commentaries.

MOSAIC GEOLOGY

When Burnet published his *Sacred Theory of the Earth* in an attempt to demonstrate that Cartesian mechanism was consistent with divine providence (see Figure 7.2), it drew strong criticism from Newtonians because it used the outdated Cartesian physics to account for the features of the earth's surface. Moreover, it did so in a way that seemed to many to support the deistic notion that while some creator God produced the world according to a clear plan, he did so in such a way that his ongoing providential activity was unnecessary (Force 1985, 38). Furthermore, Burnet argued that the Mosaic story of creation was an allegorical account, "writ in a vulgar style and to the capacity of the people" (Force 1985, 38).

In 1692, the Deist Charles Blount, printed excerpts from Burnet that seemed open to anti-scriptural interpretation in *The Oracles of Reason*. Soon after, Whiston, who was particularly concerned with the spread of deistic ideas, began his work, which would demonstrate that a Newtonian account of the origin of the earth would be far more supportive of the Mosaic text and would thus be capable of undermining Deist attempts to use natural theological arguments to attack revealed religion.

Opposed to allegorical interpretations of Scripture, like his mentor Newton, Whiston insisted that the Mosaic story was no mere fable, but rather "an Historical and True Representation of the formation of our single Earth out of a confus'd Chaos, and of the successive and visible changes each day, till it became the habitation of Mankind" (Force 1985, 41). Nonetheless, he allowed for some relaxation of an absolutely literal interpretation, emphasizing a kind of modified literalism: "We never forsake the plain, obvious, easie and natural sense, unless where the nature of the thing itself, parallel places, or evident reason, afford a solid and sufficient ground for so doing" (Greene 1959, 28) (see Figure 7.3). This view exactly paralleled Newton's, who had to deny that stars were created on the fourth day because that violated common sense. Instead, Newton and Whiston both argued that

A NEW

THEORY

OF THE

EARTH,

From its ORIGINAL, to the CONSUMMATION of all Things.

WHEREIN

The CREATION of the World in Six Days,
The Univerſal DELUGE,
And the General CONFLAGRATION,
As laid down in the Holy Scriptures,
Are ſhewn to be perfectly agreeable to
REASON and PHILOSOPHY.

With a large Introductory Diſcourſe concerning the Genuine Nature, Stile, and Extent of the *Moſaick* Hiſtory of the CREATION.

By *WILLIAM WHISTON*, M. A.
Chaplain to the Right Reverend Father in God,
JOHN Lord Biſhop of *NORWICH*, and
Fellow of *Clare-Hall* in *Cambridge*.

LONDON:
Printed by *R. Roberts*, for *Benj. Tooke* at the
Middle-Temple-Gate in *Fleet-ſtreet*. MDCXCVI.

Figure 7.1. Title page of William Whiston's *New Theory of the Earth*, which sought to demonstrate the compatability between Newtonian natural philosophy and a predominantly literal reading of Genesis. Courtesy of Special Collections, Honnold/Mudd Library.

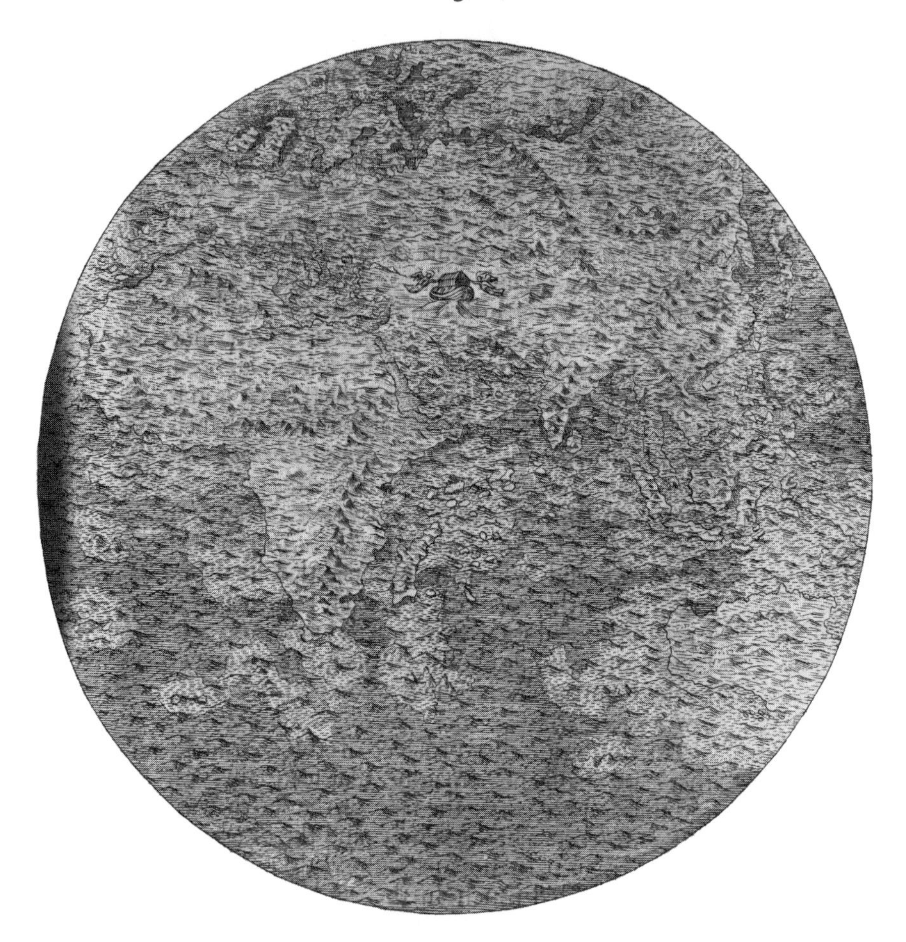

Figure 7.2. Illustration from Thomas Burnet's *Sacred Theory of the Earth* showing the Ark riding on the waves of the biblical flood that covers all of the continents. Courtesy of Special Collections, Honnold/Mudd Library.

the stars "were to be referred rather to the fourth day than any other if the air then first became clear enough for them to shine through it & so put on the appearance of lights in the firmament to enlighten the earth" (Force 1985, 52).

Given his sense that the Genesis account was a genuine historical account of creation and the early years of earth history, Whiston began by asking his readers to suppose that a comet coming into the gravitational pull of the sun was captured and drawn into an orbit around the sun at an appropriate distance from that body. The vaporous tail

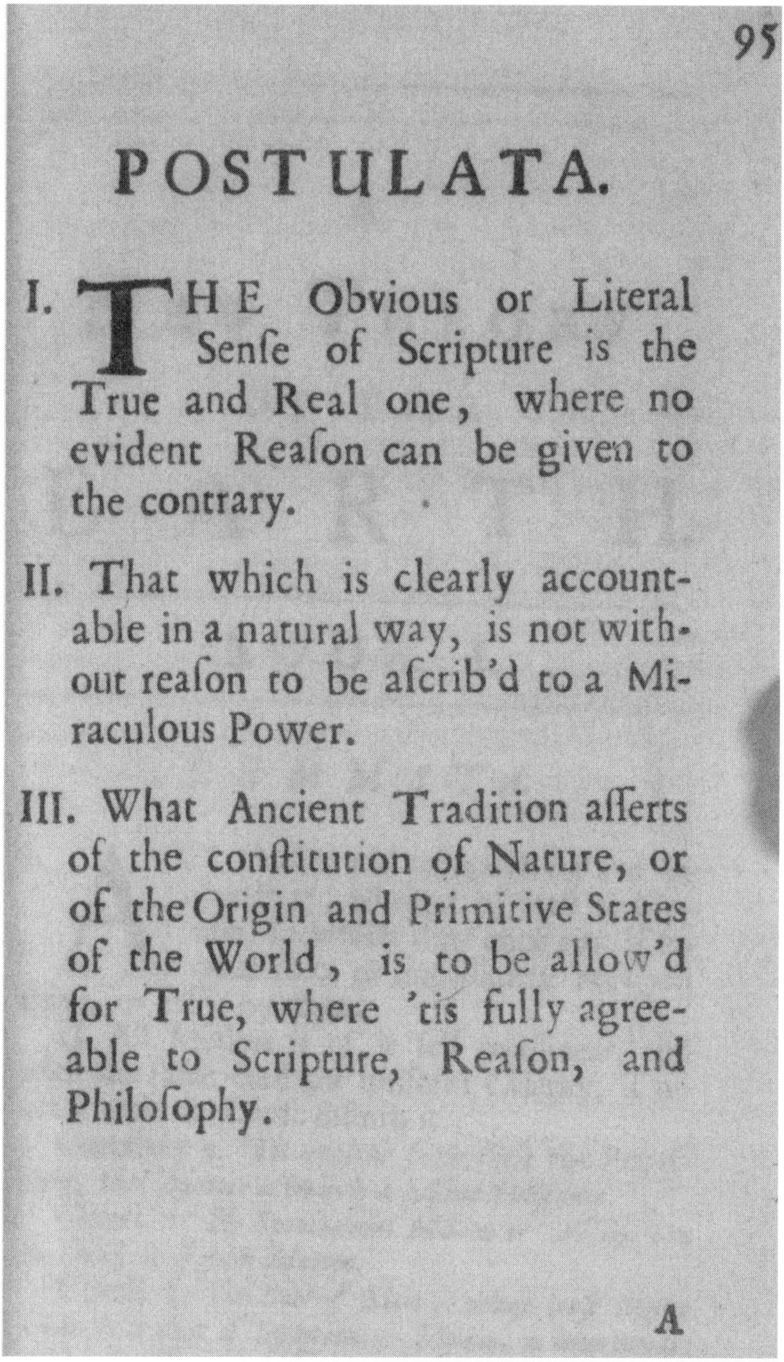

95

POSTULATA.

I. THE Obvious or Literal Senſe of Scripture is the True and Real one, where no evident Reaſon can be given to the contrary.

II. That which is clearly accountable in a natural way, is not without reaſon to be aſcrib'd to a Miraculous Power.

III. What Ancient Tradition aſſerts of the conſtitution of Nature, or of the Origin and Primitive States of the World, is to be allow'd for True, where 'tis fully agreeable to Scripture, Reaſon, and Philoſophy.

A

Figure 7.3. The set of postulates that are to guide the interpretation of Scripture according to William Whiston's *New Theory of the Earth*. Note that his is a slightly modified version of biblical literalism, which allows reason to determine when nonliteral interpretations might be necessary. Courtesy of Special Collections, Honnold/Mudd Library.

of the comet would gradually collect around the central head in roughly concentric rings of very dense fluids, water, and earth. Initially, there is no reason to expect the comet to have a daily rotation about its center. It would only have its annual revolution about the sun, so a day would be equal to a year, and the long day would be ideal for the growth of plant and animal life.

Now, suppose that a second comet passed close by the earth (see Figure 7.4). Its gravitational pull would distort the shape of the earth, causing the outer crust to crack and releasing the waters trapped inside, which would join with the waters in the tail of the passing comet (which encompassed the earth for forty days) to create a great flood. The greater pull of the passing comet on the extended near side of the earth would also produce a rotation of the earth about an axis through its center, perpendicular to the path of the comet, changing the day length to twenty-four hours and tilting the earth relative to the plane of the ecliptic (i.e., the plane in which the earth orbits around the sun) to produce the seasons as we now have them. We now have a "scientific" account of earth history—that is, one that draws only from established scientific principles—that is consistent with the historical account given by Moses, who described the phenomena without offering a scientific explanation. Indeed, by appropriately timing the passage of the planet and appropriately assigning its mass, Whiston was also able to explain the relation between the solar year and the lunar month and to demonstrate that if the timing of the flood was chosen to satisfy astronomical criteria, it was consistent with Archbishop Usher's estimate of the date of the flood (2349 B.C.E.) based on biblical chronology (Greene 1959, 30).

The tradition of biblically focused geology continued through the eighteenth century in works such as Jean André Deluc's *Lettres physiques et morales sur l'histoire de la terre et de l'homme* (1779), and into the nineteenth century in such works as John Townsend's *The Character of Moses Established for Veracity as an Historian, Recording Events from the Creation to the Deluge* (1813), and John Macculloch's *Proofs and Illustrations of the Attributes of God from the Facts and Laws of the Physical Universe* (1837).

These new works, however, tended to suggest that the formation of the earth involved a protracted process preceding the creation of humanity and that the processes involved in the initial stages of the development of the earth were radically different from the processes that continue to reshape the surface of the earth on a much more modest

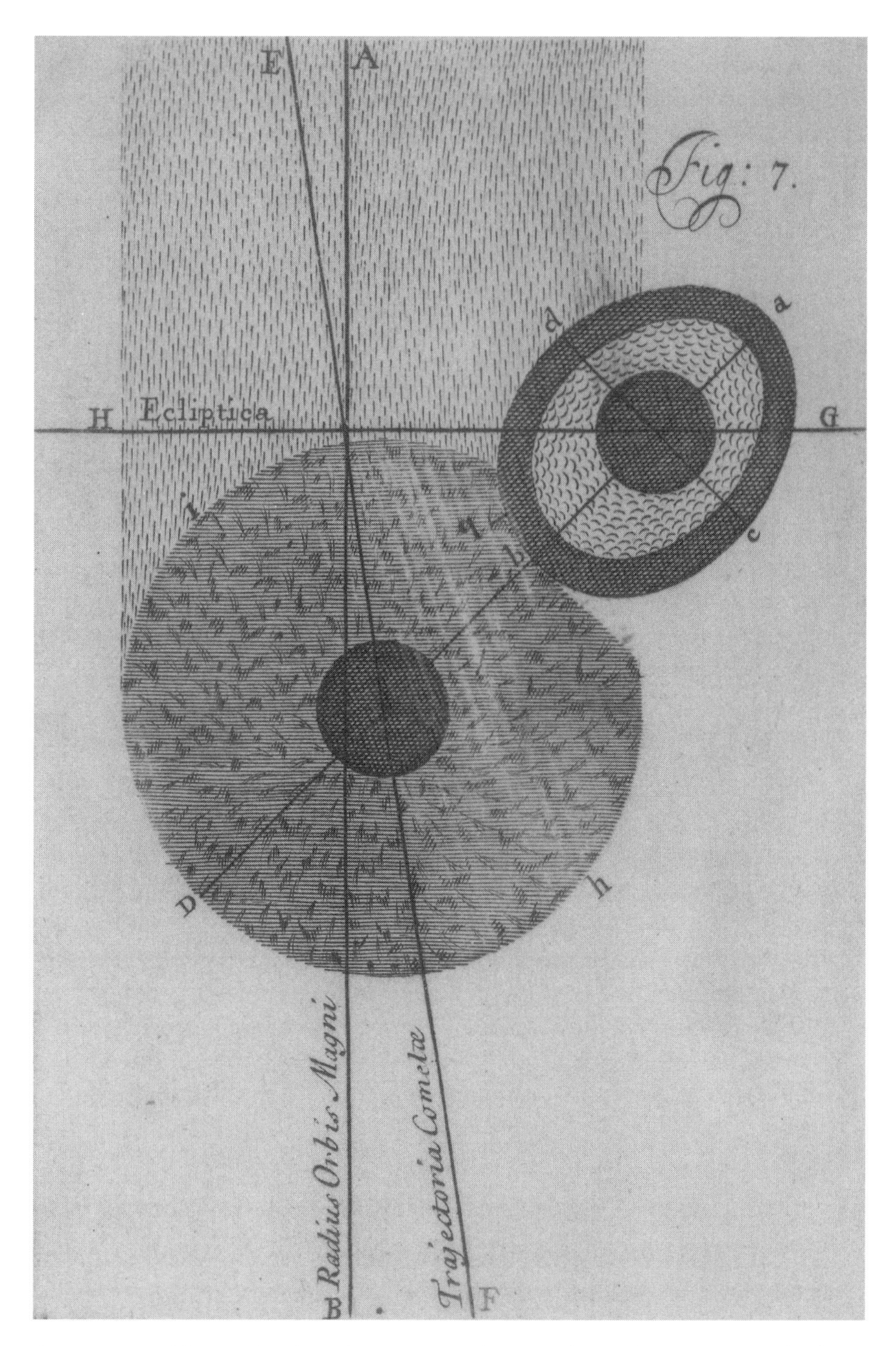

Figure 7.4. Whiston's drawing of a comet passing near the earth and creating the forty-day biblical flood as the earth is caught in the comet's liquid "tail." Courtesy of Special Collections, Honnold/Mudd Library.

scale today. According to Deluc, for example, the earlier period was dominated by the precipitation of the various rock strata out of some primordial solution at indefinitely long intervals. This was followed by a period in which the water scoured out great subterranean caverns. The roofs of some of these caverns subsequently caved in leaving great chunks upthrust or folded. As the seas gradually collected into the underground caverns, the uppermost strata and upthrust slabs from the collapse of earlier caverns began to appear above the surface of the oceans, creating a series of continents and islands. Before the emergence of the continents, fish appeared in the seas (explaining why fossils, or "fish ears," were frequently found on dry land) and were followed by the appearance of vegetation and animals. Sometime around six thousand years ago humans appeared, and then around four thousand years ago a great cataclysm—corresponding to the Mosaic flood—occurred, ushering in the present geological regime.

Different biblical geologists accounted for the great flood in different ways, but they all tended to see it as ushering in a new order. More importantly, they all tended to claim that while their geological arguments were independent of the Genesis story, they supported it in all of its essential details. Jean Deluc, whose account of the stages of the history of the earth provided a model for most that followed, insisted that his entire argument did not depend "on any reference to the book of Genesis," but it nonetheless "demonstrated the conformity of geological monuments with that supreme account of that series of operations which took place during the *Six Days,* or periods of time, recorded by the inspired penmen" (Gillispie 1959, 57, 59). From the perspective of these Mosaic geologists, any failure of geology to support a literal reading of Genesis would undermine their geology, their faith, or both, since they insisted upon both a narrowly literal reading of Scripture and the close integration of religious and scientific knowledge. In large part this argument was revived in the twentieth century in connection with present-day "creation science."

SECULAR GEOLOGY AND THE AGE OF THE EARTH

Biblically oriented interest in earth history soon interacted with new information being developed by geologists, cosmologists, anatomists, and natural historians, leading in two distinct but related and interacting directions. In one direction lay a tradition of secular scientific discussions of geological change, including James Hutton's *Theory of the Earth* (1795), Georges Cuvier's *Discours sur les révolutions de la surface du globe* (1811), and Charles Lyell's *Principles of Geology* (1830–1833). Within this tradition there was tremendous disagreement regarding appropriate methods of argument. Should one appeal only to processes experienced in the present to explain past events (uniformitarianism), or should one be allowed to infer the existence of more powerful forces from the evidences of past phenomena (catastrophism) as the biblically oriented geologists had? Similarly, there was disagreement regarding which kind of primary forces—either deposition out of water (Neptunism), or the crystallization and amorphous solidification of molten material, followed by the uplift of mountains (Vulcanism)—was most responsible for the initial formation of the earth's crust.

While secular works—especially Neptunist works and those of later catastrophists, who argued that both water and heat were critical actors but that they acted in the past on a scale no longer observed in nature—often acknowledged the support offered to religion through their results, they insisted upon the autonomy of scientific ideas and activities, rejecting both a role for the Genesis narrative in shaping or evaluating geological claims and the notion that Scripture was intended to teach about the natural world. Thus, while they were inclined to allow some dialogue and interchange among scientific and religious discourse, they saw the two domains as sufficiently separate that neither was critically dependent on the other. These works generated a vigorous response from Mosaic geologists such as Richard Kirwan, whose *Geological Essays* (1799) attacked Hutton's work most directly, and Deluc, whose *Treatise on Geology* (1809) took on all geologists who undermined belief in the literal interpretation of the Genesis account of origins.

For present purposes, two things are particularly important about early-nineteenth-century secular geologists. One is that their standards for evidentiary support for geological claims were substantially higher than those of the early Mosaic geologists. Second, there was a major divide between uniformitarians and catastrophists, which would subsequently become important for religious reasons. Uniformitarianism, as a feature of one important geological tradition, was initially articulated by the natural historian George Louis Leclerk, Comte de Buffon (1707–1788) (hereafter, Buffon), but it became the foundation of a powerful tradition in the hands of James Hutton and his friend and popularizer John Playfair, whose *Illustrations of the Huttonian Theory of the Earth* (1802) was the vehicle through which most geologists and almost all laypersons learned about Hutton's ideas. It came to dominate secular geology around 1830 after it served as the foundation upon which Charles Lyell built his three-volume *Principles of Geology* (1830–1833).

All versions of uniformitarianism began from the first of Isaac Newton's "Rules of Right Reasoning in Natural Philosophy" from the *Principia*, which stated that "We are to admit no more causes of natural things than are such as are both true and sufficient to explain their appearances" (1962, 398). In interpreting this rule, Newtonians generally understood the notion of a "true" cause to mean a cause whose existence had been previously established in some context beyond the context of the phenomenon to be explained. For example, Newton had considered gravity as a "true" cause for the motion of objects on the surface of the earth because it had been established as the cause of the motion of the planets around the sun and the moon around the earth before he attempted to use it to account for the fall of heavy bodies on the earth. In the first volume of his monumental *Natural History* (1749) (see Figure 7.5), Buffon extended this principle to earth history, arguing that we must assume that those processes that we can currently observe at work should account for future and past change:

> [T]o give consistency to our ideas, we must take the earth as it is, examine its different parts with minuteness, and, by induction, judge of the future [and past] from what at present exists. We ought not to be affected by causes which seldom act, and whose action is always sudden and violent. These have no place in the ordinary course of nature. But operations uniformly repeated, motions which succeed one another without interruption, are the causes which alone ought to be the foundation of our reasoning. (Greene 1959, 63–64)

By 1778, when he produced his most complete account of earth history, Buffon had backed away considerably from his earlier uniformitarian views in order to allow a cataclysmic event to account for the origin of the earth and to grant a much greater role to heat phenomena. But James Hutton reinstituted a uniformitarian emphasis in his *Theory of the Earth with Proofs and Illustrations* (1795), while linking it to thermally driven processes.

BUFFON'S

NATURAL HISTORY

OF

The Globe, and of Man;

BEASTS, BIRDS, FISHES, REPTILES, AND INSECTS.

CORRECTED AND ENLARGED

By JOHN WRIGHT, M.Z.S.

W. Harvey, del. J. Thompson, sc.

LONDON:

PRINTED FOR THOMAS TEGG, 73, CHEAPSIDE:

J. CUMMING, DUBLIN;

AND R. GRIFFIN AND CO. GLASGOW.

1831.

Figure 7.5. Title page from the 1831 English abridged edition of Comte de Buffon's *Natural History*. Though Buffon's *Natural History* was first published between 1749 and 1778, new editions continued to appear well into the nineteenth century. Courtesy of Special Collections, Honnold/Mudd Library.

According to Hutton, the three presently acting geological processes—the deposition of particulate matter in streams, lakes, and on the ocean floor; the consolidation, chemical alteration, and uplifting of strata under high heat in the interior of the earth (evidenced by volcanic eruptions); and the erosion of rock and land by wind and water—operate slowly and in a kind of balance with one another. That is, the latter process wears away what the first two processes builds up. Furthermore, no other processes can be hypothesized to account for past changes. So, Hutton concluded, geology offers no way to guess at the present age and the future duration of the earth: "the result of this physical inquiry is that we find no vestige of a beginning,—no prospect of an end" (Greene 1959, 86). Playfair stated the situation in a slightly more acceptable way by acknowledging that God presumably created the world and might bring it to an end. But he agreed with Hutton that, strictly speaking, scientific investigations could not reach to such creative or destructive acts as the beginning or termination of the world.

If the uniformitarians were correct, geology could offer no support to the Mosaic story of creation. In fact, it seemed to positively undermine the story of the universal flood, since while local floods were quite common, a flood on the scale of the biblical flood was beyond all present humans' experience. As a consequence, uniformitarianism was seriously opposed by Mosaic geologists. But it was also attacked by secular geologists, led by Georges Cuvier, for quite different, scientific, reasons.

In the first place, they argued that the uniformitarians' definition of "true" cause was ultimately defective, if for no other reason than the logic that there had to be some first case in which a cause was established, and in that case there could be no appeal to another context in which the cause had been previously established. To insist that every new phenomenon be explained by an established cause thus seemed quite arbitrary.

More importantly, Cuvier and his catastrophist colleagues argued that there was overwhelming observational evidence that allowed one to infer the existence of cataclysmic events with a high degree of probability. A student of the fossil contents of the various rock strata present in the Paris basin, Cuvier observed that dramatically different flora and fauna left fossil remains in successive layers of rock. The remains of fresh water species in one stratum were often replaced by either the fossilized remains of salt water creatures or dry land creatures in the next. Based solely on the nature of the fossil record, then, Cuvier argued that the sea had repeatedly washed over the region that now included Paris. Furthermore, he insisted that these inundations as well as other past geological events were frequently discontinuous:

> It is . . . extremely important to notice that these repeated inroads and retreats were by no means gradual. On the contrary, the majority of the cataclysms that produced them were sudden. This is particularly easy to demonstrate for the last one which by a double movement first engulfed and then exposed our present continents, or at least a great part of the ground which forms them. It also left in the northern countries the bodies of great quadrupeds, encased in ice and preserved with their skin, hair, and flesh down to our own times. If they had not been frozen as soon as killed, putrifaction would have decomposed the carcasses. And, on the other hand, this continual frost did not previously occupy the places where the animals were seized by the ice, for they could not have existed in such a temperature. The animals were killed, therefore, at the same instant when glacial conditions overwhelmed the countries they inhabited. This development was sudden, not gradual, and what is so clearly demonstrable for the last catastrophe is not less so of those that preceded it. (Gillispie 1959, 99–100)

Given the cataclysmic character of the great floods and freezes that punctuated the otherwise continuous operations of volcanoes, erosion, and so forth, Cuvier was also forced to infer that whole new ensembles of organisms reappeared after each catastrophe. Each ensemble was characterized by greater complexity than the last, and human fossils appeared only after the last great discontinuity.

While catastrophism, with its multiple catastrophes and multiple creations of plant and animal populations, hardly paralleled the Genesis story, it at least supported the notion of a great flood and the late creation of humankind. As a consequence, it held special appeal for such ordained Anglican geologists and natural theologians as William Buckland, who taught geology and minerology at Oxford, and Adam Sedgwick, who taught geology at Cambridge. While both distanced themselves from the more rigid Mosaic geologists, they continued to hope to demonstrate that geological knowledge was at least consistent with the essential doctrines revealed in Scripture. The cause of catastrophist geology was helped substantially by a set of discoveries reported by Buckland in 1823 in his *Reliquiae Diluvianae: or, Observations on the Organic Remains Contained in Caves, Fissures, and Diluvial Gravel, and on other Geological Phenomena, Attesting the Action of an Uni-*

versal Deluge (1823), and it found its crowning statement in Buckland's *Geology and Minerology Considered with Reference to Natural Theology* (1836). Buckland had discovered several caverns in which the bones of hyenas and what seemed to be their prey were covered in a silt layer corresponding to the last of Cuvier's deluges, and human remains were found in none, thus supporting both the existence of a far-reaching flood and the relatively late appearance of humans. Both Buckland and Sedgwick incorporated some of Hutton's thermally driven phenomena into their catastrophism. But they renounced his uniformitarian emphasis, as did most geologists during the 1820s. In this case, religious views and scientific views seemed to be mutually supportive.

Uniformitarianism was effectively revived by Charles Lyell, a Scot who had traveled to Cambridge to become one of Sedgwick's students. Lyell seems to have been just as religious as his catastrophist teacher. He was, however, more driven by a strict inductivist philosophy of science—probably because of his acquaintance with Scottish Common Sense philosophy in Edinburgh before his trip south. Lyell was fully aware of Cuvier's work and his claim that we must posit forces other than those we presently experience to account for such phenomena as glacially preserved mastodons. But Lyell argued in the first volume of his *Principles of Geology* that such phenomena were better left unexplained than explained by the assumption of forces beyond our range of experience: "When difficulties arise in interpreting the monuments of the past, I deem it more consistent with philosophical caution to refer them to our present ignorance of all the existing agents, or all their possible effects in an indefinite lapse of time, than to causes formerly in operation, but which have ceased to act" (Appleman 2001, 50).

In addition, Lyell argued that evidence accumulating since Cuvier's major work indicated much greater continuity in the fossil record than Cuvier had observed. On the one hand, there were some organisms that seemed unchanged in distant strata in a single location. More importantly, it was clear that discontinuities in the geological record occurred at different times in different places. Thus, if there were catastrophes, they were local rather than global. Furthermore, if one looked at the series of strata across the globe, some were missing in one place, while others were missing at other places. This was best explained, according to Lyell, on the assumption that erosion simply wore away some strata in some places and other strata in others, depending on local conditions. Finally, if one reconstructed a "complete" geological column from evidence gathered at many locations, there were very substantial continuities of fossils from one stratum to the next.

Lyell's work tended to draw the newer generation of geologists back toward uniformitarianism, and this meant a move away from the use of geology for natural theological purposes. Even Lyell's old teacher, Sedgwick, eventually admitted in his retirement address as President of the Geological Society of London that he could no longer discover clear evidence for the Mosaic flood in the geological record.

The early-nineteenth-century contest between uniformitarian and catastrophist geology unquestionably bore on religious interests. With rare exceptions, the more central scriptural religion was to one's concerns, the more likely one was to embrace catastrophism. But ultimately it was a scientific contest whose outcome was decided on the basis of a combination of observational evidence and commitments to particular notions of what constituted proper scientific method. Eventually, even committed natural theologians came to accept uniformitarian views.

ACCOUNTING FOR CHANGE OVER TIME

A second direction taken by eighteenth- and nineteenth-century inheritors of the new interest in earth history involved the creation of a broader tradition of natural history that increasingly sought to incorporate new domains of knowledge into a continuous story of change. This new material included understandings of change in the heavens drawn from the theoretical celestial mechanics of Immanuel Kant, Pierre Simon Laplace, and John Nichol, as well as from the observation and classification of nebulae by William Herschel. It incorporated new geological evidence that the kinds of animals and plants inhabiting the earth changed dramatically but slowly over time. It emphasized new evidence regarding the geographical distribution of plants and animals discovered by the myriads of plant, insect, bird, shell, and animal collectors stimulated by colonization and by the craze for naturalizing that had been promoted by Swedish naturalist Carl Linnaeus. It integrated new studies regarding the different "races" of humans stimulated by Linnaeus's classification scheme, a growing general interest in comparative anatomy, and by persons seeking to justify—or condemn—the practice of slavery and the imperialist exploitation of non-Europeans. It borrowed from new studies of individual growth and development at the embryological level made possible by improved microscopes and exemplified in the writings of Johannes Müller and Carl von Baer. Finally, it even included new understandings regarding how human societies "progressed" from hunting to pastoral to agricultural to commercial forms, which had been developed by Enlightenment figures such as Adam Ferguson, Adam Smith, and John Millar in Scotland, by Anne-Marie-Robert Turgot in France, and by Giambattista Vico in Italy. All of these new materials were woven into increasingly comprehensive and coherent narratives of progress.

The new comprehensive natural histories shared one major feature, regardless of whether they were formally materialist or dualist in their basic assumptions and whether their authors were nominally Christian or atheistic. Their arguments were driven by newly uncovered scientific evidence and by an obsession for inclusiveness that is matched today only by such theoretical physicists as Steven Weinberg, whose goal is to discover a Theory of Everything (TOE) that would unify the four known classes of forces: gravitational, electromagnetic, and the strong and weak. When their authors called on their theories to justify belief in a wise creator God, that justification tended to be completely unrelated to the Genesis account, and it was likely to be incidental to, rather than central to, their main argument. Finally, these works of scientific synthesis—especially those of Comte Georges-Louis Leclerc de Buffon, J. B. Lamark, and Robert Chambers—provoked intense discussions on both scientific and religious grounds. They were generally viewed as far too speculative and based on inadequate inductive evidence. They were shown to rest on incorrect interpretations of the evidence or on obsolete theories. And, increasingly in the nineteenth century, they were attacked because of their unacceptable political, social, or religious implications.

Since subsequent "conflict theorists" have often emphasized the role of religious opposition relative to scientific and political opposition to these theories, it is important to try to assess the impact of various forms of criticism to understand how important religious views were in suppressing or challenging these "scientific" works. Furthermore, only by understanding both the popularity and the hostile reactions to Lamarck and Chambers can one appreciate why Darwin developed his arguments for evolution in the way that he did and why

Darwinian evolution by natural selection seemed so threatening to some—but by no means all—religious figures.

BUFFON

Though there had been synthetic scientific accounts of the formation and character of the universe and all things in it beginning with Plato's *Timaeus* (ca. 360 B.C.E.) and the Roman Lucretius's poem *On the Nature of the Universe* (ca. 50 B.C.E.), Buffon's *Epochs de la nature* (1778) was the first to incorporate the latest results of Newtonian natural philosophy, cosmology, natural history, physiology, and ethnography in a narrative that began with the formation of the earth and ended with a discussion of the social arrangements of different races of men.

Buffon began by proposing that the earth and other planets were created when a large comet had a near collision with the sun, throwing off large globs of white hot, molten solar matter, some of which became the planets. The earth thus started out as a large, hot drop of material; and since the earth has a magnetic field, Buffon assumed that the bulk of the earth was molten iron. Given this assumption, Buffon made an estimate of the age of the earth using Newton's law of cooling, which states that the drop in temperature of a hot object in a cold environment during a short time interval is always proportional to the temperature difference. Buffon did experiments allowing heated iron balls of different sizes to cool from white heat to near room temperature, then he extrapolated his results to a ball the size of the earth, calculating that starting at the temperature of white hot molten iron, it would have reached the current temperature of the earth's crust only after 100,696 years. Allowing for the fact that a substantial portion of the earth is made up of non-ferrous, "fusible and calcareous materials," he adjusted the time to "74,047 years, approximately" (Toulmin and Goodfield 1977, 147). While this was brief, compared to the nearly infinite time assumed by Hutton shortly after, it was the first experimentally established time line to extend far beyond the biblical time scale. To minimize his obvious divergence from Christian tradition, Buffon posited six epochs of earth history corresponding to the six "days" of creation, but each lasting thousands of years.

Shortly after Buffon began his *Natural History* in 1749, he came to believe in the fixity of species, defined as groups that preserve themselves by reproduction. By the time he began the *Epochs of Nature*, however, he had become convinced that species can change over time, within relatively narrow limits. The basic structure of each family of organisms was established initially through the existence of an "internal mold" that directed the development of an infant out of the materials provided by its diet. Though Buffon did not discuss anything approaching the design argument, it is fairly clear that he thought that the internal mold was created to work best in the original environment inhabited by the organism. Suppose that an organism is created in a certain environment. Next, either the environment changes around the organism or the organism migrates into a new environment. Since the new conditions do not provide the same nutrition as its original environment, the organism's internal mold will not have the same materials to work on. As a consequence, the organism will "degenerate" within its new context into a new species, incapable of interbreeding with its parent species. If the parent species moved off into several different new environments, several related degenerate species might be formed.

Believing that humans emerged initially at the east end of the Mediterranean basin, Buffon was inclined to see those races of men that had moved away into Africa, East Asia, Oceana, and the Americas as degenerate forms. At one point, Buffon even suggested that apes might possibly be degenerate forms of humans, but he generally argued that humans were uniquely characterized by thought and speech and that "matter alone, though perfectly organized, can produce neither language nor thought, unless it be animated by a superior principle" (Greene 1959, 183).

Finally, Buffon argued that life might have spontaneously emerged within the organic world, especially within the hot and wet medium of the early oceans, so he thought it possible that relatively simple life forms originated without any special creation by a transcendent entity. This assumption was probably held by the majority of Continental scientists of his time, for it was not until the work of Louis Pasteur in the mid-nineteenth century that compelling evidence against spontaneous generation was offered. The fact that he thought that simple life could arise spontaneously, however, did not mean that Buffon believed that all life derived from spontaneously generated life forms. Especially since he generally conceived change as degenerative rather than progressive, Buffon suggested that a number of initial forms of life had been created separately at various periods in the earth's history. Some of these forms died out as the climate cooled dramatically, and the many varieties that now exist are the largely degenerate descendants of about twenty-four recently created forms.

Buffon's twenty-four-volume *Natural History*—which included the *Epochs of Nature*—was wildly successful. In an inventory of 500 private libraries in France at the end of the eighteenth century done by the literary historian Daniel Mornet, Buffon's work appeared in over 200, second only to the Abbe Pluche's work on natural theology. The best-known works of Rousseau and Voltaire, on the other hand, appeared in fifty-three or fewer. Moreover, Buffon was rapidly translated into English, going through at least three editions in London by 1791. A second ten-volume English version appeared at Edinburgh between 1807 and 1815.

LAMARCK

Buffon's protégé Jean-Baptiste Lamarck (1744–1829) offered new speculations regarding the mechanisms by which organisms change over time in works leading up to his *Philosophie zoologique* (1809). Fossil evidence showed that organisms generally became more complex and highly differentiated over time, challenging the notion that successive generations should be seen as degenerating from some superior initial form. Moreover, this evidence was consistent with a widespread belief in progress that had been building since the rise of the Christian humanist movement in the late Renaissance. As late as the mid-seventeenth century, there could be a serious debate among advocates of "the ancients" and "the moderns" regarding whether people in the ancient world—including Athens and Rome for secular scholars and the pre-lapsarian world for certain religious groups—were mentally and morally superior to those in the current European world. Jean Jacques Rousseau even continued to claim that "civilization" reflected a basic corruption of a more moral natural man well into the eighteenth century. But beginning with the Scottish followers of Montesquieu, such as Adam Ferguson, Adam Smith, and John Millar, philosophical historians began to offer a theory of material and mental progress in which societies moved from hunting and gathering as the basic modes of subsistence to herding, agriculture, and finally to commerce. Looking at the "savages" in Africa and the new world as reflecting earlier stages in the development of European societies, Europeans saw their own technologies, their own levels of knowledge and sophistication, and their own material conditions as much more advanced, and they created a notion of progress that was projected onto the natural world as well.

Lamarck posited the existence of a fundamental drive toward complexification within all matter, beginning with the simplest chemical elements, which joined to form more complex molecules. Life, then, arose spontaneously (he thought as a consequence of an electrical discharge in a gel-like soup of large inorganic chemicals). Indeed, according to Lamarck, life had arisen frequently in the course of the earth's history, and each time it formed, it formed in exactly the same way. This single, simple life form, acting at different times under the same general progressive law, then gave rise to the sequence of increasingly complex life forms, all the way up to humankind. If there had been no environmental changes over time, the current ensemble of animals would simply reflect animals at different stages along this single path of transformation from the simple to the more complex and highly differentiated. The simpler animals would represent the path that started from the more recent spontaneous origin of life, while the more complex would represent the path that had been in existence longer.

Like Buffon, Lamarck admitted that environmental conditions were different both from time to time and from place to place. Moreover, it seemed to Lamarck that the fossil remains of organisms did not lie on any simple, single scale of greater complexity; so he posited a second mechanism for change that supplemented the first. Taking a cue from the way in which human muscle groups atrophy with disuse and grow in bulk with increased use, Lamark argued that certain animals choose to work certain muscles and organs harder in order to meet their needs in a changing environment. In a famous example, he argued that under drought conditions when food was scarce, some antelope-like animal chose to stretch its neck to reach leaves that were not otherwise accessible. Now, argued Lamarck, in a crucial move, that animal was able to pass on its stretched neck to its progeny, increasing their likelihood of survival. After several generations, the antelope-like creature had been transformed into the giraffe while some of its fellow animals may have wandered off into a more moderate climate and followed their developmental path into full-fledged antelopes. The result was a "branching" evolutionary tree.

Lamarck's theories were unflinchingly materialistic in a way that Buffon's had not been. Following views being developed by Pierre Cabanis and a group of colleagues known as the Idéologues, Lamark argued that life, consciousness, and even will, were all consequences of the increasingly complex organization of matter, and that there was no need for any kind of extramaterial infusion from an outside divine agency. Lamarck thus seemed to challenge virtually all of the basic Christian ideas associated with Genesis. Not only did he deny the Mosaic flood, but he denied the fundamental difference between humans and animals—the divine infusion by which humans were made in the image and likeness of God and granted dominion over the earth and its inhabitants. Lamarck was neither an atheist nor an agnostic, but he was definitely in the Kantian camp in believing that scientific theories and religious doctrines should be kept radically separate. Lamarck's apparent irreligion, linked with his republican sentiments, made his views doubly suspect in England, even among anatomists who were becoming convinced through the work of German physiologists that change over time was a feature of organic life. Thus, Joseph Henry Green, professor of anatomy at the College of Surgeons in London, assured his students in 1840 that though he considered nature to evince "a series of evolutions from the lowest to the highest," he did not concur "in supposing that there is any power in the lower to become, or to assume the rank and privileges of, the higher, upon any such fanciful scheme as that proposed for the invertebrated animals by that laborious and otherwise meritorious naturalist, Lamarck." Instead, Green saw that ascent is "the manifestation of a higher power acting in and by nature" (Desmond 1989, 264).

THE *VESTIGES OF THE NATURAL HISTORY OF CREATION*

The pre-Darwinian culmination of comprehensive and speculative narratives of progress was unquestionably Robert Chambers' (1802–1871) anonymously published *Vestiges of the Natural History of Creation* (1844), which began with the formation of stars and concluded with the mental and moral development of humankind, all in one grand story of progressive change. Chambers' scientific interests led him to become a serious amateur naturalist and geologist and to publish a narrowly focused scientific monograph, *Ancient Sea-margins, as Memorials of Changes in the Relative Level of the Sea*, in 1848. But he was also drawn to write his grand synthesis after reading John Nichol's *Views of the Architecture of the Heavens* (1837). Nichol presented contemporary theories regarding the evolution of nebulae into stars in a highly popular form and suggested that in time it might be possible to unite "the mystical evolution of firmamental matter with the destinies of man" (Secord 2000, 92). Chambers apparently took this as his challenge.

Chambers drew heavily from Lamarck, accepting the notion that changes in the organization of matter could produce emergent properties, such as life and consciousness. But he rejected the Frenchman's emphasis on willfull change and the inheritance of acquired characteristics. Chambers argued that there was a fundamental progression that directed the embryological development of all organisms. If the embryo was stopped early in this development, it became a very simple organism; if it was allowed to go further, it might become a fish; further, it might become a reptile; still further, a bird; and if allowed to develop to the end of its progressive sequence, it would become a Caucasian male human.

Consider the short time that humans have been watching organisms reproduce. They always seem to do so in an unchanging way; but the natural law being followed may actually produce a transmutation after many, many generations, allowing a new branch of organisms to depart from the original pattern (Chambers 1994, 206–211). Turning to the progressive changes of vertebrates, Chambers invited his readers to consider a simple diagram (see Figure 7.6):

> The foetus of all the four classes may be supposed to advance in an identical condition to the point A. The fish there diverges and passes along a line apart, and peculiar to itself, to its mature state at F. The reptile, bird, and mammal, go on together to C, where the reptile diverges in like manner, and advances itself to R. The bird diverges at D, and Goes on to B. The mammal then goes forward in a straight line to the highest point of organization at M. This diagram shews only the main ramifications; but the reader must suppose minor ones, representing the subordinate differences of orders, tribes, families, genera, etc., if he wishes to extend his views to the whole varieties of being in the animal kingdom. (Chambers 1994, 212–213)

Why some fetuses only progress a certain distance along one of the many available paths, while others go on to their highest possible form, Chambers could not say for certain, though he suggested that it might be due to the forces generated by external conditions operating on the fetus.

Chambers argued that humans were, at least for the present, the most advanced animals around, although there was no reason to believe that higher species would not evolve at some future point in time. He insisted that human superiority to other animals was noth-

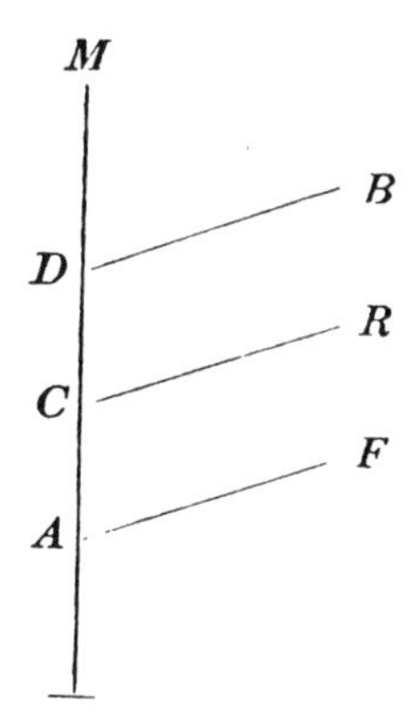

Figure 7.6. Diagram from Robert Chambers' *Vestiges of the Natural History of Creation*, illustrating evolutionary branching. (P. 212)

ing but a consequence of physical structure. "By virtue of his superior organization, his enjoyments are much higher and more varied than those of any of the lower animals," Chambers wrote (1994, 379–380). This was a mixed blessing, because the extreme complexity of the circumstances affecting a human's happiness made it likely that each human would frequently be unhappy. Fortunately, the experience that humans have in working out their problems and those of their society lead ultimately to the improvement of human reason and to a net increase in their happiness.

Understanding that his work was likely to shock the sensibilities of many, and wanting to protect his publishing house from possible scandal, Chambers arranged to have the *Vestiges of the Natural History of Creation* published anonymously. It was not until 1884 in the twelfth edition that Chambers was revealed as the author and the book was published by his own publishing house in Edinburgh, signaling that its content was no longer dangerous and scandalous.

It seems as though virtually everyone who read any non-fiction, from Victoria and Albert down to literate members of the working class, read Chambers' work. What they found there had at least the form of natural theology, for it frequently referred to the processes it described as evidences of design; but many suspected that these references were not entirely sincere.

Whatever Chambers' private religious views, he did suggest in the *Vestiges* that scientific knowledge was unable to tell us about the "first cause," or who established the laws of nature in the first place: "We advance from law to the cause of law, and ask, What is that? Whence have come all these beautiful regulations? Here science leaves us. . . . Man pauses breathless at the contemplation of a subject so much above his finite faculties." (1994, 25–26). There may be other grounds upon which belief in a first cause might rest; but they are not scientific.

While *Vestiges* was a great popular success, religious figures split dramatically over its implications, with High Church Anglicans and Catholic converts at Oxford generally welcoming what they perceived as its attack on natural theology as a central prop to religious belief, while the supporters of natural theology found it dangerous and repulsive. The response of scientific scholars was, on the other hand, overwhelmingly negative. Opposition focused on the paucity of evidence provided by Chambers for his sweeping claims and on his tendency to borrow heavily from both scientific popularizers and from scientists whose works were deemed by some reviewers to be out of date or incorrect. Adam Sedgwick undoubtedly spoke for many scientists when he insisted in his *Edinburgh Review* article of 1845 that the author of *Vestiges* had simply not paid his (or in Sedgwick's mind, more likely her) dues as a scientific investigator (Secord 2000, 243). Perhaps more pertinent in a scientific critique were Sedgwick's claims that the author's theory of development was too vague and lacking in specific information about which environmental influences produced which kind of changes. Even more telling was Sedgwick's catastrophist-induced insistence that the fossil record simply did not support the kind of continuity and constant progress suggested by Chambers' theory of development.

That Sedgwick's scientific objections were not altogether dissociated from his religious views was suggested in his review of *Vestiges,* in which he reiterated the theme of many natural theologians that the most fundamental goal of natural science was to lead us to an awareness of the character of our creator. But Sedgwick was even more adamant about his morally and religiously based distaste in his private correspondence. In April of 1845, he wrote to his former student Charles Lyell, condemning the book as "[g]ross credulity and rank infidelity joined in unlawful marriage, and breeding a deformed progeny of unnatural conclusions!" He went on to say,

> If the book be true, the labours of sober induction are in vain; religion is a lie; human law is a mass of folly and a base injustice; morality is moonshine; our labours for the black people of Africa were works of madmen; and man and

woman are only better beasts! When I read some pages of the foul book, it brought Swift's satire to my mind, and filled me with such inexpressible disgust that I threw [it] down. (Gillispie 1959, 165)

While this reaction was no doubt more extreme than most, its features were probably characteristic of the response of those liberal Anglican scientists who had attached themselves deeply to the notion that natural theology provided the most compelling evidence for belief in scriptural Christianity, and who were likely to have been leaders in the antislavery movement.

Scottish evangelical members of the Free Church of Scotland, led by Thomas Chalmers, one of the *Bridgewater Treatise* authors, generally took much the same position as the Anglican natural theologians, though they were more inclined toward a very literal reading of Genesis and to admit the miraculous intervention of God in nature. Thus, the *North British Review*, which reflected Free Church views, insisted the following in its response to *Vestiges*, which was written by the distinguished scientist David Brewster:

If it has been revealed to man theat the Almighty made him out of the dust of the earth, and breathed into his nostrils the breath of life, it is vain to tell a Christian that man was originally a speck of albumen, and passed through the stages of monads and monkeys, before he attained his present intellectual pre-eminence. (Secord 2000, 275)

For those religious persons within the High Church tradition or the English, as opposed to Scottish, evangelical traditions, which had long been suspicious of and largely opposed to natural theology—groups whose views were reinforced by the newer voices of German Protestant theology—the *Vestiges* finally seemed to demonstrate the impossibility of raising religion on the back of science. As the weekly *Guardian*, the voice of the Oxford Tractarian movement, stated:

The *Vestiges* warns us, if proof were required, of the vanity of those boasts which great men used to make, that science naturally led to religion. (It may lead beyond the experiment and the generalization, to vast theories,—visions and histories for the imagination, realities of order and law for the reason—to a *substitute* for religion.) In a world of widening and self-sustained order, an Epicurean atheism is not so difficult; something deeper than the facts of natural science is required to undercut its premises. It is the metaphysician—the abstract thinker—who is wanting in the field. (Secord 2000, 256–257)

Though the *Vestiges* may have continued to encourage many conservatives and evangelicals to avoid science altogether, others saw its clear break from scriptural religion as a license to pursue scientific knowledge, now without worrying about its religious implications. The *Guardian* expressed this central idea: "Keep in view the great principle that belief in God does not depend upon the natural—that nature is not the real basis of religion, and we can safely afford full and free scope to science" (Secord 2000, 258). Within the group associated with the Tractarians, *Vestiges* was widely admired rather than condemned. Mark Pattison, for example, placed the *Vestiges*, along with John Herschel's *Preliminary Discourse on the Study of Natural Philosophy* and Alexander von Humbolt's *Cosmos*, among the great classics of science. Henry Acland, another Tractarian, who became Reader in Anatomy, drew heavily from *Vestiges* in his lectures regarding human origins, though he insisted that God "superadded the spiritual gifts which sever him from the world beneath, and bind him to higher nature, and in common with them, to his Maker" (Secord 2000, 254–255).

On the eve of Darwin's publication of *On the Origin of Species* in 1859, then, the general topic of evolution or transmutation of species was a subject of widespread discussion. Evolutionary theories were intimately associated with the claim that humans were a product of developmental laws that made them close relatives to the great apes, and evolutionary theories were often viewed as perfectly acceptable by some conservative and evangelical religious intellectuals—who saw them as undermining traditional liberal approaches to religion. On the other hand, they were generally opposed by the traditional scientific elite in Britain, and many evangelical Scots, who were still deeply committed to natural theology. A scientific tradition that had begun largely from the desire to find scientific support for the Genesis account of the origin of the earth played a leading role in promoting a spreading commitment to the radical separation of science and religion and to deep divisions within the scientific community over the most basic aims of their activity.

Chapter 8

What to Do about Darwin?

By the time of the publication of Charles Darwin's *On the Origin of Species* in 1859, science and religion interactions in Europe and America had taken on a bewildering number of forms. Especially in Scotland and America, the longstanding tradition of natural theology maintained its vitality, now often linked with Common Sense philosophy. Though German theology was slow to take hold in England, beginning around 1860 there was a movement among liberal clergy at Oxford that insisted on the radical separation of faith from knowledge and which drew from German hermeneutic scholarship. This movement, which argued that the Bible should be approached like any other piece of literature, produced a notorious volume, *Essays and Reviews*, in 1860. *Essays and Reviews* was, however, so far from representing the clerical mainstream that over 10,900 clerics signed a declaration renouncing its views.

Most Catholics, who had seen the formation of a position accepting a completely naturalistic understanding of the world in connection with the mechanical philosophy, subject to the notion that God created the universe and was responsible for the creation and emplacement of each individual human soul in its body, had relatively little inclination to condemn Darwinian evolutionary theory, so long as it was willing to stop at the physical development of humans. This reticence was certainly due in part to the negative publicity relating to the Galileo affair. But when scholars tried to push beyond biological topics to apply evolutionary concepts to church history, theology, and doctrine, then Catholic spokespersons tended to respond by rejecting their views. Italian Catholics, in particular, were inclined to support the notion of evolution by natural selection up to the point of human mental and emotional development; but they reacted strongly against the attempts to extend evolutionary theory into a full blown form of materialism or into a religion in its own right.

Some scientists, including Thomas Henry Huxley in Britain, developed a set of ideas that generally go under the label of "scientific naturalism." Their position seemed to many to challenge the possibility of God's activity in the world entirely. The fundamental assertion of scientific naturalism is that there can be no warrant for believing any knowledge claim that does not emerge out of the methods of natural science. This position led to open conflict over such topics as the possible efficacy of prayer in dealing with weather and health. Significant numbers of intellectuals preceded or followed Huxley into what he called "agnosticism," or commitment to the proposition that there is inadequate evidence to justify a belief in God or to justify the denial of God's existence. Through such scientific-naturalist advocates of evolution as Huxley, Darwinism became associated in the minds of at least some believers with tendencies toward atheism, creating an important source of antagonism.

A number of those who were scientific naturalists, and those who were advocates of the early positivist claim that positive knowledge replaced the more primitive forms of religious and metaphysical knowledge, combined these perspectives with evolutionary notions, largely adapted from Darwin's *Descent of Man* (1871), to create new anthropological and sociological interpretations of the development of religion. Led by E. B. Tylor, Andrew Lang, and J. G. Frazer, these scholars reinvigorated and gave an evolutionary shape to the classical anthropological tradition of exploring the foundations of religion discussed in chapter 6.

In this chapter, we will explore just a sample of the diversity of science and religion interactions that characterized the middle and late nineteenth century by focusing on some of the religious responses to Darwinian evolutionary ideas in Britain and America.

For a great variety of reasons, some political, and some more broadly cultural, geography did matter in connection with science and religion interactions in general and in connection with religious responses to Darwin in particular. (Those interested in the relationships between Darwinism and religion outside of Britain and the United

States should begin by consulting Paul 1979; Kelly 1981; Kohn 1985; chapter 8 of Brooke 1991; Gregory 1992; and Numbers and Stenhouse 1999.)

THE CHARACTER OF CHARLES DARWIN'S *ON THE ORIGIN OF SPECIES*

Charles Darwin's (1809–1882) *On the Origin of Species* appeared in 1859, at a time when evolutionary ideas were being widely discussed in the aftermath of the uproar that attended Robert Chambers' *Vestiges of the Natural History of Creation*, and the appearance of Herbert Spencer's *Social Statics* (1851), which focused on the application of evolutionary ideas to social issues. In view of the strong scientific opposition to Chambers' work, Darwin took great pains to address the kind of technical questions regarding gaps in the fossil record that Adam Sedgwick had raised. In particular, he explained the complex processes by which fossils were preserved and by which successive layers of rock were deposited in any given place. He argued that it would take extremely rare circumstances to find two layers of rock containing fossils in one location that were laid down close to one another in time; so one could not expect to see a local record of continuously varying organisms. Only by constructing a complete geological column from strata collected from a number of places could one find evidence of continuous small variations. Furthermore, Darwin developed a strong argument for viewing species as nothing more than strongly marked varieties, in order to undermine the position of those who admitted natural variation and selection but denied that the very strong evidence offered for the natural selection of varieties within species justified extending the argument across the traditional boundaries between species.

Darwin was, however, impatient with the extreme anti-theoretical orientation of some critics, especially those such as Herschel, who insisted with some of the more conservative Common Sense philosophers that scientists must stick strictly to the facts, and he was disinclined to accept their criticisms. Responding to their narrow inductivist emphasis on drawing generalizations from facts alone, Darwin offered an argument very much like one developed by Auguste Comte, whose *Positive Philosophy* he undoubtedly knew through the translation by a friend and frequent visitor to his home, Harriet Martineau. Comte had insisted that facts were always theory-dependent and chosen to support whatever theory a scientific author was hoping to establish. Similarly, Darwin wrote to his friend, the political economist Henry Fawcett, in September of 1861:

> About thirty years ago there was much talk that geologists ought only to observe and not to theorize; and I well remember some one saying that at this rate a man might as well go into a gravel-pit and count the pebbles and describe the colours. How odd it is that anyone should not see that all observation must be for or against some view if it is to be of any service. (1903, 195)

This Darwinian attitude toward fact and theory made it difficult for those scientists and theologians deeply committed to inductivist philosophies of knowledge to accept the arguments of *On the Origin of Species*. For them, Darwin's need to explain why the fossil record showed very few examples of local continuous variations rejected the priority of fact over theory that they took as a central feature of scientific investigation. For them, the primary fact was local discontinuity, and no theory could be allowed to take precedence over that fact. A continuing commitment to Common Sense understandings of science among some evangelical groups underlies the creationist criticisms of Darwinian evolution even into the twenty-first century.

A second feature of Darwin's work also offended many Anglo-American and French Christian readers; for Darwin's theory of evolution by natural selection appeared to accept and promote materialism—that is, the assumption that mind, spirit, or soul is merely a consequence of the organization of matter and that the independent existence of any non-material entity is illusory. Darwin was careful not to make the materialist implications of the theory explicit in *On the Origin of Species*, in which his discussions of behavioral characteristics, instincts, and intelligence were carried out as if they might be inherited independently of bodily structures. But most readers understood Darwin as a materialist; and in *The Descent of Man and Selection in Relation to Sex*, published in 1871, he discussed at length the relationships between brain structure and mental functioning, reinforcing the notion that he was committed to the idea that mental events have somatic (bodily) causes.

The interpretation of Darwin as a thoroughgoing materialist and scientific naturalist was undoubtedly helped along by the fact that many scientists and laypersons—especially in America and on the Continent among German speakers—were introduced to Darwinian evolution through the many works of Ernst Haeckel, a German Darwinian, who frequently and aggressively expressed anti-Christian views. Darwin's apparent materialism probably helped with his acceptance in Germany, where Feuerbach's works had stimulated a materialist movement among young intellectuals. There, most scientists by midcentury were, either, like Haeckel, committed to some form of materialism, or they thoroughly divorced their religious from their scientific views. Materialism was almost certainly a liability with respect to most Anglo-American audiences, however. In much the same way, Darwin was introduced to French readers through a 1862 translation by Clémence Royer, who included an overtly anticlerical preface and set of explanatory notes, leaving largely Catholic audiences with the clear impression that Darwin was an overt atheist.

Finally, Darwin's theory of evolution diverged from any of those previously developed, except for those of the ancient atomists and his contemporary, Alfred Russell Wallace, by following a completely non-teleological pattern that Darwin had first met in his careful reading of David Hume. Darwin suggested that the initial variations in hereditary material are probably random, rather than directed toward any particular end. Evolution by natural selection is thus a process that has no particular goal (this is what it means to be non-teleological), and its results are not designed in any traditional sense. Those variations that favor the survival and reproduction of the organisms that carry them are simply "chosen" by natural selection—that is, their carriers produce larger numbers of progeny and eventually outnumber those organisms not so favored. On the other hand, if an initial variation makes an organism less likely to survive and reproduce, its progeny will be fewer and fewer in successive generations, and it will eventually die out.

Arguing explicitly against any conscious design of organisms to suit their environments, Darwin pointed out that such design was belied by evidence of several kinds. First, it was belied by the fact that when new organisms were introduced into an environment, they often drove out the native inhabitants, suggesting that the original inhabitants were not as well adapted to the environment as the newcomers. Second, when one studied the fauna of nearly identical deep cave environments in Europe and America, the animals in one place were relatively unlike those in the other; but in each locality, they were very similar to the animals that lived on the surface near by. Thirdly, many organisms included features—like the human appendix—that were either useless or positively harmful in the present, but that might have been valuable to distant ancestors. In all of these cases, if the organism was "designed," it was clearly designed imperfectly by some inferior workman.

Especially for those whose philosophical views had been influenced by Kantian or idealist philosophies, the notion that any natural process—above all, any biological process—could be understood without reference to the purpose or end that it served was simply unthinkable. Furthermore, some argued that if the traditional Christian God was to be associated with such a process, it could only be in terms of establishing the initial conditions, rather than in terms of designing particular organisms, and this position smacked of an unacceptable deism to many Christians.

On the other hand, for a few religious thinkers, the fact that variations in Darwin's theory were not to be accounted for as a consequence of God's general providence expressed through natural law suggested that the apparently accidental character of variations was a consequence of the fact that they were instead the product of a direct, unknowable, specially providential divine intervention in the natural order. The fact that few variations led to survival, while many others led to death, was perfectly consistent with the Calvinist doctrine of special election, according to which some souls are saved, while most are not.

One final issue became important in understanding religious responses to Darwin's work. That was the implication of evolution by natural selection for the origin of humankind, and especially for human mental, emotional, and moral characteristics. Natural historians had long been interested in the relationship between humans and those animals that seemed most like them. A few, including Buffon and the eccentric eighteenth-century Scottish jurist and anthropologist Lord Monboddo, had even speculated about the possibility of the descent of humans from orangutans. By placing evolution on a more respectable scientific footing, Darwin's theories seemed to make this issue much more urgent.

Darwin's notebooks and private correspondence make it very clear that he had been interested in the parallels between the mental and emotional development of humans and those of other animals from the late 1830s, when he studied the development of his first two children, "Jenny," the first baby orangutan to be exhibited in the London Zoo, and "Tommy," the first young chimpanzee to be exhibited. Moreover, his studies had convinced him of human descent from animals. Thus, he wrote in 1838:

> Let man visit Orang-outang in domestication, hear expressive whine, see its intelligence when spoken [to], as if it understood every word said; see its affection to those it knew; see its passion and rage, sulkiness and very actions of despair; let him look at savage . . . and then let him dare boast of his proud pre-eminence. . . . Man in his arrogance thinks himself a great work, worthy of the interposition of a deity. More humble and I believe *true* to consider himself created from animals. (Keynes 2002, 44)

When he wrote *Origin*, Darwin chose not to address explicitly the issue of the origins of humans, probably hoping to avoid the religious problems that this issue was bound to create. But readers assumed—correctly—that Darwin viewed humans simply as part of the natural order and of the evolutionary process. Thus, when he finally published *The Descent of Man* in 1871, though it held a few important surprises in connection with its heavy emphasis on sexual selection and what we now call group selection, it surprised no one in its basic argument that everything about humankind, including its mental and moral life, could be accounted for naturalistically and without reference to God. For those who believed in the unique relationship between humans and God, in whose image and likeness they were made, this argument was extremely difficult to accept.

Alfred Russell Wallace, who developed a theory of evolution by natural selection at the same time as Darwin, on the other hand, was a deeply religious man who insisted that the vast mental superiority of humans over their nearest relatives demands that we admit that natural selection alone is not adequate to account for that difference. He argued that we must accept a special divine intervention to account for the human mind and the moral rules that it is able to discern. Even Thomas Henry Huxley, who was widely known as a tenacious defender of Darwinian ideas, and who was a self-professed agnostic, argued for a kind of human uniqueness as early as 1863 in his *Man's Place in Nature*. Huxley wrote:

> No one is more strongly convinced than I am of the vastness of the gulf between civilized man and the brutes; or is more certain that whether *from* them or not, he is assuredly not *of* them. No one is less disposed to think lightly of the present dignity, or despairingly of the future hopes, of the only consciously intelligent denizen of this world. (Appleman 1970, 320)

Of course, Huxley thought that the unique place of humanity in the natural world was simply a consequence of its development of language and culture; but others could see language acquisition as the product of God's providence acting through secondary, natural laws. Thus it was possible for some audiences to hang on to the impression—contra Darwin—that natural selection and the special place of humanity relative to God might be consistent with one another.

Mitigating all of these troubling considerations were certain crucial passages in *On the Origin of Species* in which Darwin himself seemed to suggest that his theory was, in fact, consistent with at least some interpretations of Christianity. In the final sentence of the work, for example, Darwin used clearly biblical language when he wrote of life "having been originally breathed by the Creator into a few forms or into one" (n.d., 374). In private correspondence, Darwin later expressed regret at this language, which did not accurately reflect his own unbelief; but most readers could not know of his private views.

INITIAL ANGLO-AMERICAN RELIGIOUS RESPONSES TO DARWIN

Even among secular professional biologists, the pure theory of evolution by natural selection did not dominate evolutionary thinking until the growth of modern genetics in the early twentieth century and the establishment of what has been called the "modern synthesis" between evolution and genetics in the 1930s. Generally speaking, some version of evolution involving a combination of Lamarck-like teleology with natural selectionist emphases was widely accepted almost everywhere outside of Germany, where Darwinism did prevail among professional biologists. This general hesitancy regarding evolution by natural selection should be kept in mind as we consider specific religious responses.

Among the few traditionally religious natural historians to embrace evolution by natural selection almost immediately was the American botanist Asa Gray. Gray, a moderate evangelical, was no less committed to natural theology and notions of design than most of his colleagues. But, he argued in a series of anonymous articles published in *Atlantic Monthly* during 1860, Darwin's theory was not inconsistent with theism. In the first place, natural selection was no different from any other secondary cause chosen by God to carry out his ordinary providence. As a consequence it "would leave the doctrine of final causes, utility, and special design just where they were before" (Roberts 1988, 19). In the second place, variation received no natural explanation in Darwin's theory. In Gray's mind, "[v]ariation has been led along certain beneficial lines," (Roberts 1788, 19) and since the direction of variation was not natural, it provided evidence of a divine designer. Finally, Gray pointed to the final paragraph of the *Origin of Species*, insisting that Darwin himself had acknowledged an initial divine creative act. And, insisted Gray, one divine participation in the process established the principle of divine guidance.

James McCosh, the Scottish-born philosopher who became the president of Princeton (then the College of New Jersey) in 1868, was somewhat less enthusiastic, but he did insist in *Christianity and Positivism* (1871) that evolution, properly understood, was consistent with theism. The discovery of a convincing efficient, or secondary, cause did not preclude the existence of a final cause; so a Christian could be justified in simultaneously holding that natural selection was the efficient cause of the creation of new species and that God created and designed each living entity.

Despite the favorable responses of a few intellectuals, most Protestants found Darwinism disturbing for both religious and social reasons. Perhaps no one expressed the widespread fears about the social implications of Darwinism better than the anonymous author of the *Edinburgh Review* article that followed the publication of Darwin's *Descent of Man*:

> It is impossible to over-estimate the magnitude of the issue. If our humanity be merely the natural product of the modified faculties of the brutes, most earnest-minded men will be compelled to give up those motives by which they have attempted to live noble and virtuous lives, as founded on a mistake . . . our moral sense will turn out to be a mere developed instinct . . . and the revelation of God to us, and the hope of a future life, pleasurable daydreams invented for the good of society. If these views be true, a revolution in thought is imminent, which will shake society to its very foundations by destroying the sanctity of the conscience and religious sense. (Ellegård 1990, 100)

The initial overwhelming response to fears regarding the implications of Darwinian evolution was to try to reject the theory on the grounds that it did not meet the criteria of scientific acceptability—that it was false science or what we might call pseudoscience today. There is little doubt that the primary reasons for the searching scrutiny that Darwinian scientific arguments received were often religious; but there was widespread agreement that it would be inappropriate to appeal solely to religious arguments to discredit a scientific claim. Thus, even T.H. Huxley's nemesis, Samuel Wilberforce, wrote in the *Quarterly Review*: "We cannot . . . consent to test the truth of natural science by the Word of Revelation. But this does not make it the less important to point out on scientific grounds scientific errors, when those errors tend to limit God's glory in creation or to gainsay the revealed relations of that creation to Himself" (Ellegård 1990, 99).

Americans were particularly aggressive in challenging the scientific merits of Darwinian theory; but it is ironic that, given the American's general preference for Common Sense methodology, it was a German idealist-trained biologist, Louis Agassiz, who became the leader of the American campaign against Darwinism. According to Agassiz, every entity in the world, including every living species, was the material approximation of an idea in the mind of God. The similarities among members of large groups of animals did not suggest to Agassiz a common ancestry, but rather that they were variations on a single ideal type in the mind of God. Moreover, the patterns of geographical distribution of organisms did not represent migration, with modification, from a central location, but the repeated introduction in multiple places of organisms based on the same idea.

On one central issue, Agassiz argued that the factual evidence unambiguously favored his type-theory over Darwin's theory of descent with modification, and it was this issue that attracted the attention of most inductivist-oriented Americans. Following Cuvier, Agassiz emphasized the fact that in no single place had anyone found fossil evidence of the gradual change of species. Quite different ensembles of organisms appeared in neighboring strata, as if each layer of rock represented a new burst of divine creativity. In order to incorporate these facts into his theory, Darwin was forced to offer a complicated and speculative account of why neighboring rock strata in any given place rarely represented consecutive time periods. We have already seen that this claim did not bother Darwin. In fact, as complaints about the speculative or hypothetical character of the *Origin of Species* piled up, a frustrated Darwin wrote to his friend Joseph Hooker, "I have always looked at Natural Selection *as an hypothesis* which, if it explains several large classes of facts would deserve to be ranked as a theory deserving acceptance" (Roberts 1988, 41).

If it did not bother Darwin, however, the speculative or hypothetical status of natural selection did bother a large number of scientists and religious thinkers alike. Charles Hodge spoke for many when he insisted in his *What Is Darwinism?* of 1874 that it was nothing but atheism grounded in a hypothesis; and Randolph S. Foster, the Methodist president of Drew University, concurred in an 1872 essay, "Origin of Species: Examination of Darwinism," when he insisted that Darwinism failed as science because "science wastes no time in stammering and muttering of conjectures and possibilities; that is the method of doubt, not of knowledge" (Roberts 1988, 42). Even many who admitted a significant role for hypotheses in the sciences were inclined to reject hypotheses, like that of natural selection, which seemed to challenge common religious views. Thus Sir David Brewster, a Scottish physicist, educator, and co-founder of the British Association for the Advancement of Science, reflected in 1862 on Darwin's views that:

> There are many cases indeed, in the history of science, where speculations, like those of Kepler, have led to great discoveries. . . . It is otherwise, however with speculations which trench upon sacred ground, and which run counter to the universal convictions of mankind, poisoning the fountains of science, and the serenity of the Christian World. (Ellegård 1990, 100)

Critics brought up in the tradition of natural theology were also irritated by the way that Darwin used the analogy between artificial selection by humans and natural selection. To them, artificial selection was the work of intelligent agents who always had some end in mind as they chose to breed certain animals rather than others. How, then, could Darwin claim that its successes supported the existence of the unintelligent and undirected process of natural selection? This complaint seemed doubly valid when even Darwin admitted that artificially selected varieties tended to revert to parent forms when left in the wild; so it seemed that the artificial selection process could not even create stable varieties. By the same token, some religious intellectuals were inclined to accept the notion that there was a clear line of demarcation between varieties and species. These persons admitted that Darwin might have adduced convincing evidence for the creation of varieties by both artificial and natural selection. But they contended that he had been unable show that species were created by either process.

DARWINISM AND CONCERNS ABOUT SCIENTIFIC NATURALISM

Church attendance was falling off in all classes in virtually all of Western Europe during the mid-nineteenth century, and many secular and religious scholars alike were inclined to place the responsibility or blame for this state of affairs on the growing intellectual authority of science and the aggressively anti-clerical views of advocates of scientific naturalism, such as Huxley. This blame was largely misplaced; but it was invited by some of Huxley's particularly inflammatory pronouncements, including the following statement made in 1860:

> In this nineteenth century, as at the dawn of modern physical science, the cosmology of the semi-barbarous Hebrew is the incubus of the philosopher and the opprobrium of the orthodox. Who shall number the patient and earnest seekers after truth, from the days of Galileo until now, whose lives have been embittered and their good name blasted by the mistaken zeal of Bibliolaters? Who shall count the host of weaker men whose sense of truth has been destroyed in the attempt to harmonize impossibilities—whose life has been wasted in the attempt to force the generous new wine of science into the old bottles of Judaism, compelled by the outcry of the same strong party?
>
> It is true that if philosophers have suffered, their cause has been amply avenged. Extinguished theologians lie about the cradle of every science as the strangled snakes beside that of Hercules; and history records that whenever science and orthodoxy have been fairly opposed, the latter has been forced to retire from the lists, bleeding and crushed, if not annihilated; scotched, if not slain. But orthodoxy is the Bourbon of the world of thought. It learns not, neither can it forget; and though, at present, bewildered and afraid to move, it is as willing as ever to insist that the first chapter of Genesis contains the beginning and end of sound science, and to visit, with such petty thunderbolts as its half-paralyzed hands can hurl, those who refuse to degrade nature to the level of primitive Judaism. (Moore 1979, 60)

Open hostilities had broken out between Huxley and the Anglican establishment when Huxley and the bishop of Oxford, Samuel Wilberforce, confronted one another at a session of the meeting of the British Association for the Advancement of Science on June 30, 1860. In front of an audience of over 700 men and women who were overwhelmingly anti-Darwinian, Wilberforce had turned to Huxley and, according to a magazine report, "begged to know, was it through his grandfather or his grandmother that he claimed his descent from a monkey." Huxley replied to great cheers that, while he would not be ashamed to have a monkey for an ancestor, he would "be ashamed to be connected with a man who used his great gifts to obscure the truth [as Wilberforce did]" (Irvine 1955, 6–7).

Antagonism among Christians toward Huxley—and by extension, toward Darwinism—intensified over the next several years in response to Huxley's 1863 *Evidence as to Man's Place in Nature*, which denied that the special character of humans was a consequence of any intervention of the divine in evolutionary processes and which implicitly promoted the doctrine of scientific naturalism—the doctrine that no claim that could not be established using the methods of science should be accepted as valid. Huxley admitted that humans were unique in their capacity for moral and ethical behavior and in their capacity to consciously intervene to shift the direction of evolution. But he insisted that this uniqueness was a natural consequence of the human development of language and that there was therefore no reason to accept any claim of divine intervention in the evolutionary process to produce humans.

Most Protestant thinkers, on the other hand, insisted that even if the physiological differences between humans and other organisms were minimal, this did not support the notion of common descent. Appealing to Agassiz's notions, they pointed out that at most the similarities implied that God had a small number of ideas in mind in creating physical organisms. Generally, they argued that the human ability to reason abstractly, and the human belief in spiritual agencies evidenced in religious rituals attested to for all known human cultures, so distinguished humans from other animals that they warranted the conviction that some supernatural agency was responsible.

Darwin published *The Descent of Man* in 1871, offering an evolutionary account for the birth and survival of religious institutions. Religions provided support for the development of altruistic behavior, which in turn made those early societies in which religions developed more cohesive in tribal warfare and therefore, more likely to survive. From the perspective of many religious thinkers, this argument suggested that Darwinism was intrinsically implicated in the anti-Christian views associated with scientific naturalism; for Darwinism, like Feuerbach's materialism, sought to interpret religion as a product of human activity selected for by natural forces rather than as a form of worship mandated by an independently existing supernatural Divine.

Toward the end of the century, E. B. Tylor's anthropological investigations of religious practices—which combined a positivist sense of progress with an evolutionary perspective—seemed to confirm the antagonism between Christianity and evolutionary theory. Tylor purported to demonstrate that such Christian rituals as baptism and consecration were mere "survivals" or "relics" of savage practices (Burrow 1970, 256). Moreover, he seemed to send a challenge to advocates of traditional religions in the second volume of his *Anthropology*, published in 1881:

> Unless a religion can hold its own in the front of science and morals, it may only gradually, in the course of ages, lose its place in the nation, but all the power of statecraft and all the wealth of the temples will not save it from eventually yielding to a belief that takes in *higher* knowledge and teaches *better* life. (Burrow 1970, 258)

It was difficult for any religious person to acknowledge the idea that scientific knowledge was higher than religious belief or that a secular life was in any way better than a religious one.

Within a few years after Huxley's first inflammatory essays, the American chemist-turned-historian William Draper, who had been the featured speaker at the session where Wilberforce and Huxley were commentators, published *A History of the Conflict between Religion and Science* (1874), and a battle that continues almost unabated even today within a small segment of secular and religious intellectuals was well and truly joined. It was often the case that the "conflict" between science and religion was largely a consequence of deep misunderstandings. This was particularly the case in connection with one of Huxley's close friends, the Irish physicist John Tyndall.

Many opponents of scientific naturalism lumped Tyndall with Huxley as a materialist and scientific naturalist for easily understood but ultimately unjustifiable reasons. In fact, Tyndall's religious views had been formulated under the influence of Kantian and German idealist philosophies while he was preparing for his Ph.D. in natural philosophy at Marberg. He accepted the Kantian distinction between the deterministic and knowable realm of phenomena dealt with by science and the ultimately unknowable, but immensely important, realm of feeling and will that he understood to be the proper domain of religion. Tyndall was perfectly willing to admit that scientific naturalism was an inadequate philosophy for the necessities of a truly human life. He insisted, however, that for purposes of dealing with the physical—as opposed to the psychic—world, materialism and scientific naturalism were adequate and that religious claims were illegitimately considered. As a consequence, he became embroiled in a debate that angered the religious community in Britain and America.

When Bishop Wilberforce and several of his Anglican colleagues called in August of 1860 for a national day of prayer to stop the wet weather that threatened harvests throughout Britain, Tyndall saw an opportunity to demonstrate the lack of connection between religion and physical phenomena. Prayer might well allow a person to prepare him- or herself to accept a set of circumstances; but, according to Tyndall, it could have no impact on such a natural phenomenon as the weather. As a consequence, Tyndall proposed an experiment that had been suggested by a friend to test the physical efficacy of prayer.

A group of believers would pray for several years on behalf of patients in a specified hospital ward, unbeknownst to the patients; then the mortality rate for that ward would be compared to the mortality rates for wards for which no special prayers had been said. A significantly lower mortality rate in the ward for which prayers were offered would indicate the efficacy of prayer. The failure to discover a significant difference would confirm Tyndall's claim that God does not respond to prayer with respect to physical phenomena. This so-called prayer gauge test was not carried out (though similar tests were done with ambiguous results in the late twentieth century) because religious thinkers argued that it would be unfair. Some argued that God would surely not be willing to submit to such a test because he required faith. Others argued that healing particular individuals might conflict with some higher purpose of God's. Still others argued that God's ordinary providence was already being carried out through natural laws and that it was unreasonable to expect a specially providential intervention for any but a major divine priority.

Tyndall's reputation as an arrogant scientific naturalist, comparable in his disdain for religion with Huxley, was cemented in 1874 by his retiring presidential address to the British Association for the Advancement of Science. Up to this point, he had insisted upon the independence of science and religion—though he saw and approved the fact that the social authority of science was growing at the expense of religious authority. But he concluded his 1874 address with a direct claim that had remained quietly implicit in his earlier writings: that while scientific knowledge must not be subject to religious constraint, religious beliefs should be subject to scientific constraints. What was even worse, he did so by insisting that natural selection should be applied to the contemporary Christian religious situation:

> All religious theories, schemes, and systems, which embrace notions of cosmogony, or which otherwise reach into the domain of science, must, insofar as they do this, submit to the control of science, and relinquish all thoughts of controlling it. . . . Every system which would escape the fate of an organism too rigid to adjust itself to its environment must be plastic to the extent that the growth of knowledge demands. (2000, 382)

Tyndall's claims, coming from the presidency of the British Association for the Advancement of Science, produced a flood of angry responses from clerics and laypersons alike. Many who had strongly supported scientific activity as long as they believed that it would support religious ends now withdrew their support of what they perceived as a heartless, soulless, and anti-Christian science. Such a perception was amazingly widespread—in spite of the fact that most distinguished British scientists continued to be strong believers—in part because Huxley and Tyndall were so visible, so vocal, and such polished speakers and prose stylists.

The views expressed by the conservative Unitarian Frances Cobbe is representative of the changing attitudes of many non-scientist believers. Through the 1860s she had viewed science positively and in religious terms. Writing for *Frazer's Magazine* in 1863, she said: "Physical science, the knowledge of God's material creation, is in a sense a holy thing—the revelation of God's power, wisdom [and] love through the universe of inorganic matter and organic life" (French 1975, 366). But in the aftermath of Huxley's and Tyndall's defenses of scientific naturalism she reassessed her views. Thus, in *The Study of Physiology as a Branch of Education* written in 1883, she warned that "the moralist may discern the production of a certain type of hardness and arrogance which has developed itself in the ranks of science ever since science has ceased to be to its followers the highway to religion" (French 1975, 367). By 1889, Cobbe was advocating a full-fledged religious war against the advocates of scientific naturalism. In *The Churches and Moral Questions* she wrote:

It would truly seem that blindness has fallen on the eyes of the clergy that they do not perceive that the science which they shrink from even seeming to oppose or discredit, is the same Godless, inhuman science which teaches that there is no Father in Heaven but only an "unknown and unknowable" Author of a world of struggle and pain, unguided by a single ray of hope beyond the grave. Can['t] they recognize that this science . . . is their own irreconcilable enemy? Do they not foresee that if such science swell a little higher and sweep over England, it will swallow up their Churches and their Faith. (French 1975, 370)

ANGLO-AMERICAN PROTESTANT RESPONSES TO DARWIN AFTER 1875

By the middle of the decade of the 1870s, almost all serious naturalists had accepted some form of transformism or evolution. Furthermore, almost all agreement that natural selection played some role, though they were far from agreed on the mechanisms by which variations occurred. A widespread slogan claimed that natural selection accounts for the preservation but not the origination of forms in nature (Roberts 1988, 85). In particular, very few scientists, whether religious or not, were willing to follow Darwin into a complete abandonment of teleology until well into the twentieth century. This unwillingness was certainly promoted in 1868 when William Thomson, Lord Kelvin, calculated the age of the earth, creating an updated version of Buffon's cooling rate calculations. According to Kelvin, if the earth had no internal sources of heat (an assumption shown to be wrong only in the early twentieth century), it could have taken no more than 100 million years to cool from a molten state to its present temperature. Though it was hard to estimate how long the mechanism of natural selection alone working on random variations might have taken to produce the variety of organisms on earth, no one believed that it could have happened in such a short time. Thus, many evolutionists were inclined to argue that variations must have been directed in some way to speed up the process.

With the professional triumph of evolutionary theories, most Anglo-American Protestants turned away from rejecting evolution on purportedly scientific grounds. Moreover, a substantial majority of Christians refused to take part in the battle against evolution, which linked evolutionary theories to scientific naturalism. Instead, they sought to harmonize scientific and religious views by adopting carefully crafted interpretations of both evolution and Christianity so as to make them work together. There were very few elements common to all the harmonizers of the late nineteenth century except for a sense that both religion and science would benefit from being allies rather than rivals and a rejection of Darwin and Huxley's anti-teleological emphasis. There were advocates of both progressive evolution and of degenerative evolution among religious thinkers, but virtually all agreed that if Darwin were correct that there was no direction to evolution, then religion would lose virtually all of its evidence from the natural world.

Given their generally teleological tendencies, religious thinkers could then view evolution—even evolution that depended heavily on natural selection—as simply "the method by which the Divine had populated the world with organisms adapted to the conditions of their existence" (Roberts 1988, 122). A few religious theists, including George F. Wright and James McCosh, actually welcomed the inefficient character of natural selection on the grounds that was consistent with the doctrine of the election of the favored few. Thus argued Wright, "If Calvinism is a foe to sentimentalism in Theology, so is Darwinism in natural history" (Roberts 1988, 132). Others, such as John Fiske and Joseph LeConte, who was much more of an optimist, argued that natural selection offered an explanation of "the dark riddle of physical evil" (Roberts 1988, 134), by indicating why evil, in the form of the death of many varieties and species, was the cost that had to be paid for a greater good—the progress explicit in most evolutionary theories.

A very few theologians appropriated the notion of a degenerative evolution, which was prevalent among French and Italian thinkers but also present in England in the late nineteenth century, to understand the doctrine of the Fall. Others, who insisted upon the centrality of the Fall for all of Christian theology, emphasized the fact that evolutionary theory was at least consistent with the notion that a preference for sin was probably transmitted from one generation to the next. Most evolutionary theists, on the other hand, agreed with Lamarck; Herbert Spencer, who extended evolutionary ideas to social progress; and the Congregational minister J. L. Diman, who argued in his 1880 Lowell Lectures in Boston that "what strikes us most forcibly about the natural world is not simply the fact of development, but the fact that this development has been progressive" (Roberts 1988, 122). Such men often had a difficult time reconciling evolution with the traditional

Christian doctrine of the Fall of man. Henry Ward Beecher, for example, saw progressive evolution as a way around what he considered to be the "repulsive, unreasonable, immoral, and demoralizing" doctrine of the Fall (Roberts 1988, 196). And Minot Savage, a Congregationalist minister-turned-Unitarian welcomed evolutionary theory on the grounds that it proved that "humanity has never fallen" (Roberts 1988, 195). This is just one important topic on which evolutionary theory seemed to some progressive theologians to invite desirable modification of traditional interpretations of Scripture.

From the perspective of what one might call "liberal" Christianity today, one of the most interesting responses to evolution incorporated this notion that scientific knowledge might reasonably force changes in traditional readings of Scripture, integrating it with the German higher criticism. This response promoted the idea that revelation itself and the religion grounded in revelation were part of a progressive evolutionary process. George Trumbull Ladd, who was a Congregational minister before teaching moral philosophy at Yale, insisted in *The Doctrine of Sacred Scriptures* (1883) that religious knowledge is "the result of a process of unfolding," and that "it is given to us in the form of development" (Roberts 1988, 160). Similarly, Elizabeth Stuart Phelps argued that "what we call inspiration is a growth. It unfolds with history and like history. It is subject to evolution, like the race" (Roberts 1988, 161). David Hill, president of Lewisburg University, wrote that because "all nature reveals development, it would be anomalous indeed, if we did not find it in a revelation designed for men of different attainments and different consequent needs" (Roberts 1988, 161).

In its most extreme form, such a doctrine transformed the fundamental role of Scripture in religion. Totally rejecting notions of biblical inerrancy and infallibility, such liberal clergy as David N. Beach, a Congregational minister in the Midwest argued that the Bible was simply a starting place for "the blazing of a path out toward the never-to-be-overtaken horizon of enlightenment and growth" (Roberts 1988, 164). This trend reached its nineteenth-century culmination in Lyman Abbott's *The Evolution of Christianity* (1892), which offered a contemporary Christianity without heaven, hell, original sin, or a divine Christ.

For many nineteenth- and early- twentieth-century liberal religious thinkers, especially those whose primary concerns were with social justice, the cruelty and inefficiency of the process of evolution by natural selection seemed inconsistent with their understandings of a caring God or their hopes for secular progress. They could often accept one of the "progressive" evolutionary theories of Lamarck, Chambers, or Spencer; but there was something disturbing about the seemingly accidental character of Darwin's theory.

The widespread concern that rigid natural selectionism deprived life of meaning strongly shaped American scientists' views of evolution, so that most American evolutionists were insistent on presenting evolution as teleologically directed. Asa Gray had initiated this trend in his 1860 review of *On the Origin of Species*, and it was continued by such staunch evolutionists as Joseph LeConte, who wrote in his "Evolution in Relation to Materialism" of 1881 that "[i]n evolution we reach the one infinite, all embracing design, stretching across infinite space and infinite time, which includes and predetermines and absorbs every possible separate design" (Roberts 1988, 125). As a consequence, most American religious figures were relatively comfortable with evolution, though not with extreme selectionism.

In spite of the accommodations made by most scientific advocates of evolution and by many religious figures, there did remain a relatively small but extremely important minority of religiously conservative groups in both America and England—many of them strongly influenced by the Newton/Whiston ideas of Scriptural interpretation discussed in chapter 5—who insisted that the Bible should be read literally wherever possible and that its claims were inerrant. For such persons, even teleological versions of evolution seemed unacceptable. Evolution of all stripes seemed to undermine the biblical claims regarding the unique relationship of God to humankind; evolutionary attempts to account for morality seemed to undermine the notion that moral laws were the immutable consequence of the divine will; progressive evolutionary theories seemed to counter the doctrine of the Fall; and all evolutionary theories implied an age for the earth that seemed completely inconsistent with scriptural chronology, which placed the age of the earth at less than 10,000 years rather than the billions of years implied by evolutionary theories or even the 100 million years suggested in 1868 by Lord Kelvin.

In the late nineteenth century, Methodist clergy and theologians tended to lead those who argued that there was a fundamental conflict between evolutionary theories and the Bible. Thus, the Methodist theologian Miner Raymond wrote in his *Systematic Theology* of 1877 that, "if the origin of the race be found anywhere else than in that creation of a special pair, from whom all others have descended, then is the whole Bible a misleading and unintelligible book" (Roberts 1988, 212). Moreover, it was clear that he had greater confidence in the Bible than in science. This view was shared by most Seventh-Day Adventists, and even by some conservative Presbyterians, especially in the American South. The chemistry professor-turned-Presbyterian clergyman George Armstrong spoke for this group in 1888, writing: "where there is a conflict between truth or doctrine clearly taught in Scripture, and the generally accepted conclusions of science, sound logic requires that we accept the former, and reject the latter. God cannot err; science may err, in the present, as it often has in the past" (Roberts 1988, 219). Such conservative views constituted a reservoir of anti-Darwinian sentiment that could be tapped after World War I by those appalled by both the local and the international impact of Social Darwinist doctrines to produce the more widespread and virulent anti-evolutionary movements of the twentieth century.

ANGLO-AMERICAN CATHOLIC AND JEWISH RESPONSES TO EVOLUTION

Though Catholic priests and lay authors in nineteenth-century England and America offered a significant range of attitudes toward evolution, that range tended to be less favorable to Darwinian ideas than those of the liberal Protestants, but more favorable than those of most evangelical Protestants. At the same time, it was generally more favorable to Darwinism than that of French Catholics, whose enthusiasm for the views of the homegrown Cuvier slowed the acceptance of evolutionary ideas. But it was less favorable than the views of Italian lay Catholics, whose doctrinal commitments tended to be more superficial.

Everywhere, the complete acceptance of Darwinian evolution by natural selection was blocked by two crucial doctrines that had persisted within Catholicism since the sixteenth and seventeenth centuries. First was the insistence upon a teleological approach to natural knowledge. Teleology was a central feature of the doctrines of Thomas Aquinas, and an emphasis on purpose continued within the neo-scholastic, Thomist philosophy that remained the core of Catholic doctrines and of Catholic higher education into the twentieth century. Second was the additional Thomist insistence that the human rational soul was different from the merely vegetable and animal souls of all other living organisms and that each rational soul was directly implanted in a human being by God. This view had been emphasized explicitly in connection with Catholic versions of the mechanical philosophy of the seventeenth century, and it too remained a guiding principle into the twentieth century. Given these two constraints, it was at least permissible for Catholics to accept what came to be called "moderate" evolution in the twentieth century—that is, the claim that all species of plants and animals other than man were the natural product of a development from a few initial types created by God. As stated in the *Catholic Encyclopaedic Dictionary* in 1931:

> Catholics are free to believe in moderate evolution, excluding the evolution of man. Animals, as distinguished from man, are devoid of reason. Hence the animal soul, i.e., the principle which gives an animal life, is essentially material. . . . Man's soul, though depending on material things for its activities, being essentially spiritual, the evolution of man *as a whole* from the lower animals is impossible. (Allenby 1999, 173)

There was also some encouragement from Thomism to try to find an accommodation between Catholic doctrine and evolutionary theory once evolution was widely accepted among scientists; for Thomist doctrine held that there can never be a real conflict between God's Word and God's Works. When there was an apparent conflict, either the interpretation of one or the other must be revised. At least some Catholics thus followed John Zahm, a Holy Cross priest and professor of physics and chemistry at the University of Notre Dame, in accepting a version of theistic teleological evolution. Zahm was probably more popular in Italy than in the United States, and, as we shall see, he was asked by the Holy Office in Rome to stop publication of his *Evolution and Dogma* of 1896. Nonetheless, he was the acknowledged scientific leader of a relatively small and progressive "Americanist" collection of priests who sought to make room for science in Catholic culture and who even had some sympathy for German techniques of biblical exegesis. The majority of American Catholic clergy, however, feared what would happen if ordinary Catholics were exposed to such modern arguments as those of scientific naturalists like Huxley. They appealed to Pope Leo XIII, who, in 1899, sent an apostolic letter to Cardinal Gibbons of Baltimore, one of the leaders of the Americanist faction, warning its members against the potential dangers of the new biblical criticism and of scientific investigation.

Both the Americanists and the conservatives were quite opposed to full-fledged Darwinism because of its social and moral implications. But they tended to focus their attacks on Spencer's social theories and on Huxley's scientific naturalism; so there was relatively little explicit antagonism to the biological doctrine of evolution presented in *On the Origin of Species*. Even Zahm, however, attacked the more expansive views in *The Descent of Man*.

One consideration that shaped American Catholic responses to Darwin grew out of the fact that Catholicism in America was primarily a religion of recent working-class immigrants from Ireland, Italy, and Poland. In the 1880s anti-Catholic sentiment began to flare up, creating such "nativist" organizations as the American Protective Association, and the Immigration Restriction League, which argued for discriminatory "eugenic" policies grounded in Social Darwinist arguments. These groups promoted a wide variety of policies ranging from the imposition of an English language literacy test for immigrants, through the establishment of intelligence tests to screen out potential undesirables, to the involuntary sterilization of the "feebleminded," who came disproportionately from recent immigrant groups. In response to the purported Darwinian foundations of the arguments made by such groups, many Catholics felt impelled to reject both Social Darwinism and its presumed foundations. The eugenics movement did not become powerful until the early twentieth century, but the first involuntary sterilization bill was considered by the Michigan legislature as early as 1897, and Catholic opposition began to appear in the last decade of the nineteenth century.

Those who were theologically conservative often expressed their opposition by claiming that the idea of the evolution of humans developed by Darwin in *The Descent of Man* was opposed directly to the teachings of the church fathers, including Saint Thomas. Those who were more progressive tended to reject the authority of Aristotle and Saint Thomas, arguing: "Medieval armor will not turn a bullet from a modern rifle, nor will the authority of a Medieval philosopher be secure behind which to fight a modern evolutionist" (Allenby 1999, 182). Nonetheless, even the progressives rejected Darwin's non-teleological version of evolution in favor of a theistic one, drawing from the English Catholic George Mivart's 1871 *On the Genesis of Species*, which proposed that natural selection was subordinated to special powers employed by God to direct organisms into those forms that he had ordained. From such a perspective, the social "survival of the fittest" doctrines linked to Darwin's name could be abandoned while some progressive evolutionary scheme was preserved.

Zahm, like a modern-day Galileo, argued that theistic evolution was consistent with the writings of Saint Augustine and insisted that even Pope Leo XIII had admitted that teaching about the natural world was not the purpose of Scripture. So he insisted that, in any event, he was therefore not bound to any theory of either evolution or special creation except to the extent that there was scientific evidence for or against the doctrine. A few months after the appearance of his *Evolution and Dogma,* which expressed these views in 1896, Zahm was transferred to Rome, where his theistic evolution was generally frowned upon. In 1897, he was asked to cease publication on evolutionary topics; and in 1898, he was told that his *Evolution and Dogma* had been banned by the Congregation of the Index. Progressive Americanist attitudes toward science in general and evolution in particular would not finally prevail within Catholicism until the 1950s.

Just as attitudes toward theistic evolution provide an interesting entry into the tensions between progressive "Americanizing" Catholics and more traditional Catholics, so too do attitudes toward evolution reflect a significant tension between progressive Reform Jews, more moderate Reform Jews, and traditionalist Jews in America during the last quarter of the nineteenth century. Prior to 1870, American Jewish authors wrote little or nothing about evolution, though a London-based group, the Jewish Association for the Diffusion of Religious Knowledge, indicated its opposition by publishing pamphlets supporting Paley's natural theology in 1861 and 1863.

In the first few years following the publication of *The Descent of Man,* responses to Darwin became more frequent; but almost all American Jewish responses to Darwin were hostile and were focused on claiming the uniqueness of humans, especially in connection with morality and intelligence. One exception, that would signal important subsequent trends, was a frequently republished sermon by German-born Reform Rabbi Bernhard Felsenthal, who drew from the Kantian tradition to insist that because the foundations of religion lay in conscience, feeling, and intuition, rather than in phenomenal knowledge, Judaism should grant complete freedom of scientific inquiry to scientists. Furthermore, he insisted that the works of Darwin, Huxley, and others who explored the origin of species, could hold no threat for Judaism (Swetlitz 1999, 215).

This situation changed in 1874 when the radical Reform Rabbi Kaufman Kohler gave a series of sermons on science and religion, subsequently published as *Das neue Wissen und der alte Glaube!* (*New Science and Old Beliefs*). In these sermons, he reiterated Felsenthal's claim that Judaism had nothing to fear from the sciences; but he went on to accept aspects of higher biblical criticism, insisting that the importance of Genesis lay in its spirit rather than in any specific claims it made about the origins of the world or of humankind. Indeed, he insisted that Genesis was constructed out of "popular legends and poems" from the time of its composition. He went on to argue that evolution, or "the natural law of progressive development of life under favorable conditions," was well established by evidence from a variety of fields. Finally, and most importantly, Kohler claimed that evolutionary theory supported Reform Judaism by promoting a doctrine of progressive revelation—that is, by arguing in favor of progress in morality and in understanding of the Divine. This progressive revelation in turn justified remodeling Jewish ritual and theology in the light of modern science (Swetlitz 1999, 218).

Within a few months, several Reform rabbis, including Moritz Ellinger, editor of *The Jewish Times*, Rabbi Solomon Sonneschein, editor of *Die Deborah*, and Rabbi Jacob Mayer of Cleveland, who had previously opposed evolution, changed their views and supported Kohler's ideas. All began to argue that Reform Judaism was superior to both traditionalist Judaism and to Christianity as a religion suited to a scientific age. In their most extreme form, exemplified in Rabbi Joseph Krauskopf's *Evolution and Judaism* (1887), the radical Reformist's appeals to evolution verged on creating a virtual religion of science, with God defined as "an active, never changing law which shapes all matter" (Swetlitz 1999, 228).

There were certainly continuing Reform opponents of evolution, led by Rabbi Isaac Mayer Wise, whose work *The Cosmic God* (1876) offered a doctrine of periodic progressive creations that combined an ecclectic mix of Cuvier's geology and German Romantic *Naturphilosophie*. By the early twentieth century, however, theistic evolutionary theory and Reform Judaism had become closely linked.

Most traditionalist Jews in America rejected evolution. In fact, several, including Rabbi Alexander Kohut, who became a leader among traditionalist scholars, and Rabbi Abraham de Sola, who was a leader in the Montreal Natural History Society, turned to literalist Christian Bible interpreters and argued that humans had existed for only 5,740 years. De Sola promoted his Christian colleague J. W. Dawson, who offered a creationist interpretation of Genesis in *The Origin of the World* (1877). The Methodist minister Thomas Mitchell, who insisted that the world was created in six 24-hour days, was even invited to write a series of articles for *Menorah Monthly* in 1887.

There was, however, a small cadre of medically and biologically trained traditionalist American Jews who not only supported evolutionary theory, but who found a way to use Darwinian arguments from *The Descent of Man* to argue in favor of maintaining traditional Jewish law and ritual and Jewish race purity. This group, which included the Ernst Haeckel student Frederick de Sola Mendes, were all in their twenties when they established a new traditionalist and pro-evolutionary journal, *The American Hebrew,* in 1879. They argued that Darwin had clearly shown the value of religious ritual in promoting group cohesion. As a consequence, they insisted that to abandon or transform the laws and ceremonies that had helped preserve the Jewish people over centuries in hostile environments would be to give in to "social anarchy" and to undermine the greatest strength of Judaism.

This group also drew from Haeckel's doctrines and from anthropological perspectives developed in the Jewish medical community in continental Europe to argue that the very survival of the Jews depended on maintaining racial purity. "The law of fittest surviving, aided by the breeding of hereditary qualities in a pure race has given Jews a physiological and mental superiority which can be perpetuated only by the perpetuation of the race purity," they insisted (Swetlitz 1999, 227). It is ironic that these same doctrines were soon to be appropriated by Hitler and the Nazi party in Germany to promote Aryan race purity and to commit genocide on the German Jewish population.

CONCLUSION

This book began with a discussion of Galilean astronomy and Christianity and ended with a discussion of Christianity and Darwinian evolution. These two cases have long stood as the most notorious episodes in the supposed ongoing conflict between science and religion; but even in these cases it should have become clear that the stories are vastly more complicated than Draper, White, and their followers would have us believe. It is true that in each case there were loud religious voices opposing new scientific developments; but it is also true that there were other religious voices supporting them and promoting them for a variety of reasons.

In both cases, scientific resources were mobilized by some religious groups in order to support pre-existing struggles within the religious domain. Galileo's support of Copernican doctrines in the face of Catholic—especially Dominican—opposition, for example, was used by some Protestants to promote the idea that every individual has the right to interpret Scripture independent of Church authorities. It was used by others to undermine claims regarding the necessity of an ecclesiastical hierarchy that had been justified in terms of the old understandings of a celestial hierarchy in which God resided at the maximum distance from a corrupt earth situated at the bottom of the universe. Darwinian evolutionary arguments were similarly appropriated by liberal elements within Catholicism, Protestantism, and Judaism to support what they viewed as progressive changes in religious practices and doctrines. The mismatch between evolutionary accounts of the origins of the universe and of humankind, for example, was used by many religious liberals to challenge the very possibility of a literalist interpretation of Scripture or to undermine the long-standing doctrine of original sin.

It was also the case that evolutionary theories were shaped and interpreted by numerous Christian scientists to make them conform to the demands of their own faith traditions, liberal Christian biologists opting for theistic versions of evolution, which often denied Darwin's anti-teleological emphasis, while more conservative Christians saw in the extinction of vastly more species than survived a parallel to the Calvinist doctrine of election.

Even when Christian scholars opposed Galileo's version of Copernican astronomy or Darwin's version of evolutionary theory, their opposition was sometimes only incidentally or partially religious. There were strong reasons associated with disagreements within the community of scientists regarding the proper methods for generating and evaluating the legitimacy of scientific knowledge that led scientists—independent of their religious commitments—to criticize both Galileo's work and that of Darwin. Christian scientists thus sometimes had legitimate non-religious reasons for their opposition; but they not surprisingly also chose to adopt those versions of scientific method, which allowed them to reconcile their science and their religion.

While one can make a strong case that Christian demands for calendar reform played a significant role in directing money and effort toward both observational and mathematical astronomy in early modern Europe, it would be hard to claim that religious demands or motives drove evolutionary thinking during the nineteenth century. But evolutionary ideas did emerge out of a geological tradition that received a very strong push from seventeenth- and eighteenth-century attempts to provide scientific accounts of earth history that would support the Genesis account of creation.

When we move away from the Galileo and Darwin cases, we find evidence of a richer variety of interactions between the natural sciences and religious developments from the time of Copernicus to the end of the nineteenth century. There can be little doubt, for example, that the shift from Aristotelian, contemplative, theoretical sciences to more utilitarian experimental sciences, was strongly promoted by Christian humanist doctrines. Nor is there any significant doubt that the flowering of scientific activity in seventeenth-century England was strongly encouraged by religious developments, with the occult sciences such as alchemy finding support primarily among radical Protestants and experimental mechanical philosophy being strongly promoted among Latitudinarian Anglicans. The intimate connection between science and theology in early modern Europe can be seen with particular clarity in the careers of some of the most important scientists of the seventeenth century, including those of Marin Mersenne, Pierre Gassendi, Robert Boyle, and Isaac Newton.

High Church Anglicans and some evangelical groups resisted the liberal enthusiasm for natural theology that dominated religious discourse in England through the early nineteenth century because they feared that it would undermine traditional features of scripturally based Christianity, and it seems that their fears were, to some degree, justified. Eighteenth-century and nineteenth-century trends in the sciences seemed to be promoting heterodox religious views, including Deism, Free Thought, and even such scientistic religions as Auguste Comte's Religion of Humanity. Once again, however, the story is more complex than some early analysts suggested. The movement of intellectuals away from traditional scriptural Christianity was at least partially a response of the failure of established churches to live up to the promises of social justice that the powerful humanist movement had emphasized. First in France, then in Germany and Britain, disaffection with Christian churches that were more likely to promote violent conflict over minor doctrinal differences, or to support the interests of the wealthy and powerful rather than to serve the needs of the poor and powerless, led first to the rejection of established Christian churches and then to the adoption of science as an alternative source of value. This pattern was initiated in the works of Enlightenment figures such as Claude-Adrien Helvetius. It continued in the works of German theology students such Ludwig Feuerbach and David Strauss, and it culminated in the mid-to-late nineteenth century formation of such secular religious traditions as the Religion of Humanity, Marxism, and Monism—a religious movement that denied the separability of matter and spirit and that promoted scientific naturalism as a foundation for ethics—initiated by the biologist Ernst Haeckel and the physical chemist and expert in thermodynamics Wilhelm Ostwald.

Overall, John Brooke was certainly correct when he argued that science and religion interactions in Europe between 1500 and 1900 were extremely complex—sometimes mutually supportive, sometimes mutually antagonistic, and more often simultaneously supportive and antagonistic, depending on what particular place one occupied within the spectrum of both religious and scientific attitudes, ideas, and practices.

Primary Sources

The following sources have been chosen because they illuminate particularly important attitudes or arguments and because they are unlikely to be widely available in modestly sized public or school libraries. Equally important selections from such authors as Galileo, Isaac Newton, William Payley, Thomas Henry Huxley, and Andrew Dickson White have not been included because they are frequently reprinted or anthologized and are usually very easy to find elsewhere. An attempt has been made to avoid duplicating primary sources throughout this series; so if a topic that is of interest is also covered in another volume, you should refer to the primary sources in that volume to complement those presented here.

Though modern type fonts are used here, I have left the original spelling, punctuation, and paragraphing to provide a sense of the author's style. Where I have used editions other than the first edition of a work, it is simply because that edition was available to me, whereas an earlier edition would have been difficult to get. Permission to quote from the *Works of Robert Boyle* and from Thomas Burnet's *Sacred Theory of the Earth* comes from the Special Collections at the Honnold-Mudd Library of the Claremont Colleges. Permission to quote from John Ray's *Wisdom of God Manifested in the Works of Creation* comes from the Library of the Claremont School of Theology.

— 1 —

Hermes Trismagistus, *Hermetica*, vol. 1, trans. Sir Walter Scott (Oxford: The Clarendon Press, 1924 [first Latin Printing, 1471])

The following selections were taken from the *Hermetic Corpus*. Though actually written in the fourth century of the common era, these works were believed to pre-date both Moses and Plato when they became known widely in Europe as a result of Marcilio Ficino's Latin translation beginning in 1463. Combining elements from Plato's *Timaeus*, the Genesis account of creation, Gnostic doctrines regarding the evil inherent in matter, and astrological lore, the Hermetic works played a major role in shifting Christian scholars' interests away from the mere contemplation of God's works to active engagement in trying to perfect the world. In the process, they stimulated interest in astrology, astronomy, and applied sciences such as alchemy. The selections are from books 1 and 5 of *Hermetica.*

Once upon a time, when I had begun to think about the things that are, and my thoughts had soared high aloft, while my bodily senses had been put under restrained sleep,—yet not such sleep as that of men weighed down by fullness of food or by bodily weariness,—methought there came to me a Being of vast and boundless magnitude, who called me by my name, and said to me, "What do you wish to hear and see, and to learn and come to know by thought?" "Who are you?" I said. "I," said he, "am Poimandres, the Mind of the Sovereign." "I would fain learn," said I, "the things that are, and understand their nature, and get knowledge of God. These," I said, "are the things of which I wish to hear." He answered, "I know what you wish, for indeed I am with you everywhere; keep in mind all that you desire to learn, and I will teach you."

When he had thus spoken, forthwith all things changed in aspect before me, and were opened out in a moment. And I beheld a boundless view; all was changed to light, a mild and joyous light; and I marveled when I saw it. And in a little while, there had come to be in one part a downward-tending darkness, terrible and grim. . . . And thereafter I saw the darkness changing into a watery substance, which was unspeakably tossed about, and gave forth smoke, as from fire; and I heard it making an indescribable sound of lamentation; for there was sent forth from it an inarticulate cry. But from the Light came forth a holy Word, which took its stand upon the watery substance; and methoughts this Word was the voice of the Light.

And Poimandres spoke for me to hear, and said to me, "Do you understand the meaning of what you have seen?" "Tell me its meaning" I said, "and I shall know." "That Light," he said, "is I, even Mind, the first God, who was before the watery substance which appeared out of the darkness; and the Word which came forth out of the Light is the Son of God." "How so?" said I. "Learn my meaning," said he, "by looking at what you yourself have in you; for in you too the Word is son, and the mind is father of the word. They are not separate one from the other; for life is the union of Word and mind." Said I, "For this I thank you."

"Now fix your thought upon the light," he said, "and learn to know it." And when he had thus spoken, he gazed upon me eye to eye, so that I trembled at his aspect. And why I raised my head again, I saw in my mind that the Light consisted of innumerable powers, and had come to be an ordered world, but a world without bounds. This I perceived in thought, seing it by reason of the word which Poimandres had spoken to me. And when I was amazed, he spoke again, and said to me, "You have seen in your mind the archetypal form, which is prior to the beginning of things, and is limitless." Thus spoke Poimandres to me.

"But tell me," said I, "whence did the elements of nature come into being?" He answered, "They issued from God's Purpose, which beheld that beauteous world and copied it. The watery substance, having received the Word, was fashioned into an ordered world, the elements being separated out from kit; and from the elements came forth the brood of living creatures. Fire unmixed lept forth from the watery substance, and rose up aloft; the fire was light and keen, and active. And therewith the air too, being light, followed the fire, and mounted up until it reached the fire, parting from earth and water; so that it seemed that the air was suspended from the fire. And the fire was encompassed by a mighty power; and was held fast and stood firm. But earth and water remained in their own place, mingled together . . . but they were kept in motion, by reason of the breath-like Word which moved upon the face of the water.

"And the first Mind,—that Mind which is Life and Light,—being bisexual, gave birth to another Mind, a Maker of Things; and this second Mind made out of fire and air seven Administrators [the planets] who encompass with their orbits the world perceived by sense; and their administration is called Destiny.

"And forthwith the Word of God lept up from the downward-tending elements of nature to the pure body which had been made, and was united with Mind, the Maker; for the Word was of one substance with that Mind. And the downward-tending elements were left devoid of reason, so as to be mere matter.

"And Mind and Maker worked together with the Word, and encompassing the orbits of the Administrators, and whirling them round with a rushing movement, set circling the bodies he had made, and let them revolve, traveling from no fixed starting point to no determined goal; for their revolution begins where it ends.

"And Nature, even as Mind the Maker willed, brought forth from the downward-tending elements, animals devoid of reason; for she no longer had with her the Word. The air brought forth birds and the water, fishes—earth and water had by this time been separated from one another—and the earth brought forth four-footed creatures and creeping things, beasts wild and tame.

"But Mind the Father of all, he who is Life and Light, gave birth to Man, a Being like to Himself. And he took delight in Man, as being his own offspring; for Man was very goodly to look on, bearing the likeness of his Father. With good reason, then, did God take delight in Man; for it was God's own form that God took delight in. And God delivered over to man all things that had been made.

"And Man took station in the Maker's sphere, and observed all the things made by his brother, who was set over the region of fire; *and having observed the Maker's creation in the region of fire, he willed to make things for his own part also*; and his Father gave permission . . . having in himself all the working of the Administrators; and the Administrators took delight in him, and each of them gave to him a share of his own nature.

"And having learned to know the being of the Administrators, and received a share of their nature, he willed to break through the bounding circles of their orbits; and he looked down through the structure of the heavens, having broken through the sphere, and showed to downward-tending Nature the beautiful form of God. And Nature, seeing the beauty of the form of God, smiled with insatiate love of Man, showing the reflection of that most beautiful form in the water, and its shadow on the earth. And he, seeing this form, a form like to his own, in earth and water, loved it, and willed to dwell there. And the deed followed close on the design; and he took up his abode in matter devoid of reason. And Nature, when she had got him with whom she was in love, wrapped him in her clasp, and they were mingled in one; for they were in love with one another.

"And that is why man, unlike all other living creatures upon earth is twofold. He is mortal by reason of his body; he is immortal by reason of the Man of eternal substance. He is immortal, and has all things in his power; yet he suffers the lot of a mortal, being subject to Destiny. He is exalted above the structure of the heavens; yet he is born a slave of Destiny. . . .

"[God ordained the] births of men, and bade mankind increase and multiply abundantly. And he implants each soul in flesh by means of the gods who circle in the heavens. And to this end did He make men, that they might contemplate heaven, and have dominion over all things under heaven, and that they might come to know God's power, and witness nature's workings, and that they might mark what things are good, and discern the diverse natures of things good and bad, and invent all manner of cunning arts."

— 2 —

Richard Hooker, *The Laws of Ecclesiastical Polity in Eight Books*, book 1, ed. R.W. Church, 2nd edition (Oxford: Oxford University Press, 1876 [1593 original]), 4–5, 8, 9, 13–14, 15–17, 41–42, 43–44, 50–51, 78–79, 83, 86–87, 90–91

Hooker's defense of the Anglican Church against Puritan and Catholic critics emphasized the role of human reason and natural law in religion, arguing that Scripture unaided by nature was inadequate as a path to salvation. It thus stimulated a special concern with natural theology among Anglicans. In the passages collected here, Hooker explains the purpose of his work, discusses the relation between divine law and natural law, and therefore the relationship between God and nature. In addition, he articulates the claim that both revelation and natural knowledge, but not church tradition, are necessary for salvation.

The laws of the Church, whereby for so many ages together we have been guided in the exercise of Christian religion and the service of the true God, our rites, customs, and orders of Ecclesiastical government, are called into question; we are accused as men that will not have Christ Jesus to rule over them, but have willfully cast his statutes behind their backs, hating to be reformed, and made subject unto the sceptre of his discipline. Behold therefore we offer the laws whereby we live unto the general trial and judgement of the whole world; heartily beseeching almighty God, whom we desire to serve according to his own will, that both we and others (all kinds of partial affection being clean laid aside) may have eyes to see, and hearts to embrace, the things that in his sight are most acceptable.

And because the point about which we strive is the quality of our laws, our first entrance hereinto cannot better be made, than with consideration of the nature of law in general, and of that law which giveth life unto all the rest . . . namely the law whereby the Eternal himself doth work. Proceeding from hence to the law, first of nature, then of scripture, we shall have the easier access unto those things which come after to be debated, concerning the particular cause and question which we have in hand. . . .

The particular drift of every act proceeding externally from God we are not able to discern, and cannot always give the proper and certain reason of his works. Howbeit undoubtedly a proper and certain reason there is of every finite work of God, inasmuch as there is a law imposed upon it. . . . They err therefore who think that of the will of God to do this or that, there is no reason besides his will. Many times no reason is known to us; but that there is no reason thereof, I judge it most unreasonable to imagine. . . .

The works of nature are no less exact, than if she did both behold and study how to express some absolute shape or mirror always before her; yea, such her dexterity and skill appeareth, that no intellectual creature in the world were able by capacity to do that which nature doth without capacity and knowledge; it cannot be but nature hath some director of infinite knowledge to guide her in all her ways. Who the guide of nature, but only the God of nature? *In him we live, move, and are* (Acts xvii, 28). Those things which nature is said to do, are by divine art performed, using nature as an instrument; nor is there any such art or knowledge divine in nature working, but in the guide of nature's work. . . .

This world's first creation, and the preservation of things created, what is it but only so far forth a manifestation by execution, what the eternal law of God is concerning things natural? And as it cometh to pass in a kingdom rightly ordered, that after a law is once published; it presently takes effect far and wide, all states framing themselves thereunto; even so let us think it fareth in the natural course of the world: since the time that God did first proclaim the edicts of his law upon it, heaven and earth have hearkened unto his voice and their labor hath been to do his will: He *made a law for the rain* (Job xxviii, 26); He gave his *decree unto the sea, that the waters should not pass his commandment* (Jer. v, 22). . . . See we not plainly that obedience of creatures unto the law of nature is the stay of the whole world? . . . We see then, how nature itself teacheth laws and statutes to live by. The laws which have been hitherto mentioned do bind men absolutely as they are men, although they have never any settled fellowship, never any solemn agreement amongst themselves what to do or not to do. . . .

The works of nature are all behoveful, beautiful, without superfluity or defect; . . . framed according to that which the law of reason teacheth. Secondly, those laws are investigable by reason, without the help of revelation, supernatural and divine. Finally, in such sort are they investigable, that the knowledge of them is general. . . . Law rational, which men commonly use to call the law of nature, meaning thereby the law which human nature knoweth itself in reason universally bound unto, which also for that cause may be termed most fitly the law of reason; this law, I say, comprehendeth all those things which men know by the light of their natural understanding. . . .

Concerning faith, the principle object whereof is that eternal verity which hath discovered the treasures of hidden wisdom in Christ; concerning hope, the highest object whereof is that everlasting goodness which in Christ doth quicken the dead; concerning charity, the final object whereof is that incomprehensible beauty which shineth in the countenance of Christ the son of the living God . . . concerning that faith, hope, and charity, without which there can be no salvation, was there ever any mention made saving only in that law which God himself hath from heaven revealed? There is not in the world a syllable muttered with certain truth concerning any of these three, more than hath been supernaturally received from the mouth of the eternal God.

Laws therefore concerning these things are supernatural, both in respect of the manner of delivering them, which is divine; and also in regard of the things delivered, which are such as have not in nature any cause from which they flow, but were by the voluntary appointment of God ordained besides the course of nature, to rectify nature's obliquity withal.

When supernatural duties are necessarily exacted, natural are not rejected as needless. The law of God therefore is, though principally delivered for instruction in the one, yet fraught even with the precepts of the other also. . . .

When the question . . . is whether we be now to seek for any revealed law of God otherwhere than only in the sacred scripture; whether we do now stand bound in the sight of God to yield to traditions urged by the church of Rome the same obedience and reverence we do to his written law, honouring equally and adoring both as divine: our answer is no. . . .

Albeit scripture do profess to contain in it all things that are necessary unto salvation; yet the meaning cannot be simply of all things which are necessary, but all things that are necessary in some certain kind or form; as all things which are necessary, and either could not at all or could not easily be known by the light of natural discourse. . . . There is in scripture therefore no defect, but that any man, what place or calling soever he hold in the Church of God, may have by the light of his natural understanding so perfected, that the one being relieved by the other, there can want no part of needful instruction unto any good work which God himself requireth, be it natural or supernatural, belonging simply unto men as men, or unto men as they are united in whatsoever kind of society. *It sufficeth therefore that nature and scripture do serve in such full sort, that they both jointly, and not severally either of them, be so complete, that unto everlasting felicity we need not the knowledge of anything more than these two may furnish our minds with on all sides;* and therefore they which add traditions, as a part of supernatural, necessary truth, have not the truth, but are in error (emphasis mine).

— 3 —

Robert Boyle, "A Free Inquiry into the Vulgarly Conceived Notion of Nature," in *The Works of the Honourable Robert Boyle. In Six Volumes,* vol. 5 (London: 1772 [1686 original]), 162–164

Here the great English chemist, religious writer, and convert to the mechanical philosophy offers his account of the religious reasons for preferring the new mechanical philosophy to the natural philosophies of the Aristotelians and the Paracelsans—reasons widely shared by moderate Protestant natural philosophers.

First, it seems to distract from the honour of the great author and governor of the world, that men should ascribe most of the admirable things, that are to be met within it but to a certain nature, which themselves do not well know what to make of. It is true, that many do confess, that this nature is a thing of his establishing, and subordinate to him: but, though many confess it, when they are asked whether they do or no? Yet, besides that many seldom or never lifted up their eyes to any higher cause, he that takes notice of their way of ascribing things to nature, may easily discern, that, whatever their words sometimes be, the agency of God is little taken notice of in their thoughts: and however, it does not a little darken the excellency of the divine management of things, that, when a strange thing is to be accounted for, men so often have recourse to nature, and think she must extraordinarily interpose to bring such things about; whereas it much more tends to illustrate God's wisdom to have framed things at first, that there can seldom or never need any extraordinary interposition of his power. And as it more recommends the skill of the engineer to contrive an elaborate engine so as there should need nothing to the reach his ends in it but the contrivance of parts devoid of understanding, than if it were necessary, that ever and anon a direct servant should be employed to concur notably to the operations of this or that part, or to hinder the engine from being out of order; so it more sets off the wisdom of God in the fabric of the universe, that he can make so vast a machine perform all those many things, which he designed it should, by the mere contrivance of brute matter managed by certain laws of local motion and upheld by his ordinary and general concourse, than if he employed from time to time an intelligent overseer, such as nature is fancied to be, to regulate, assist, and control the motions of the part: in confirmation of which you may remember, that the later poets justly reprehend their predecessors for want of skill in the plots of their plays, because they often suffered things to be reduced to that pass, that they were fain to bring some deity upon the stage to help them out.

And let me tell you freely, that though I will not say, that *Aristotle* meant the mischief his doctrine did, yet I am apt to think, that the grand enemy of God's glory made great use of *Aristotle's* authority and errors to detract from it.

For as *Aristotle*, by introducing the opinion of the eternity of the world, (whereof he owns himself to have been the first broacher) did at least, in almost all men's opinions, openly deny God the production of the world; so, by ascribing the admirable works of God to what he calls nature, he tacitly denies him the government of the world: which opinion, if you judge severe, I shall not, at more leisure, refuse to acquaint you, (in a distinct paper) why I take diverse of *Aristotle's* opinions relating to religion to be more unfriendly, not to say pernicious, to it, than those several other heathen philosophers.

And here give me leave to prevent an objection, that some may make, as if to deny the received notion of nature, a man must also deny providence, of which nature is the grand instrument. For, in the first place, my opinion hinders me not at all from acknowledging God to be the author of the universe, and the continual purveyor and upholder of it; which is much more than the peripatetic hypothesis, which (as we were saying) makes the world eternal, will allow its embracers to admit: and those things, which the school-philosophers ascribe to the agency of nature interposing according to emergencies, I ascribe to the wisdom of God in the first fabric of the universe, which he so admirably contrived, that if he but continue his ordinary and general concourse, there will be no necessity of extraordinary interpositions, which may reduce him to seem, as it were, to play aftergames; all those exigencies, upon whose account philosophers and physicians seem to have devised what they call nature, being foreseen and provided for in the first fabric of the world; so that mere matter, so ordered, shall, in such and such conjunctures or circumstances, do all, that philosophers ascribe on such occasions to their almost omniscient nature, without any knowledge of what it does, or acting otherwise than according to the catholic laws of motion. And methinks the difference betwixt their opinion of God's agency in the world, and that, which I would propose, may be somewhat adumbrated by saying that they seem to imagine the world to be after the nature of a puppet, whose contrivance may indeed be very artificial, but yet is such, that almost every particular motion the artificer is fain (be drawing sometimes one wire or string, sometimes another) to guide and oft-times over-rule the actions of the engine; whereas, according to us, it is like a rare clock, such as that at Strassburgh, where all things are so skillfully contrived that, the engine being once set a moving, all things proceed, according to the artificer's first design, and the motions of the little statues, that at such hours performs these or those things, do not require, like those of puppets, the peculiar interposing of the artificer, or of any intelligent agent employed by him, but performing their functions upon particular occasions, by virtue of the general and primitive contrivance of the whole engine. The modern *Aristotelians* and other philosophers would not be taxed as injurious to providence, though they now ascribe to the ordinary course of nature, those regular motions of the planets, that *Aristotle*, and most of his followers (and among them the Christian schoolmen) did formerly ascribe to the particular guidance of intelligent and immaterial beings, which they assigned to the motives of the celestial orbs. And when I consider, how many things, that seem anomalies to us, do frequently enough happen in the world, I think it is more consonant to the respect that we owe to divine providence, to conceive that God is a most free, as well as a most wise agent, and may in many things have ends unknown to us, he very well foresaw and thought fit, that such seeming anomalies should come to pass, since he made them (as is evident in the eclipses of the sun and moon) the genuine consequences of the order he was pleased to settle in the world; by whose laws the grand agents in the universe were impowered [*sic*] and determined to act, according to the respective natures he had given them and the course of things was allowed to run on, though that would infer the happening of seeming anomalies and things really repugnant to the good or welfare of particular portions of the universe: this, I say, I think to be a notion more respectful to divine providence, than to imagine, as we commonly do, that God has appointed an intelligent and powerful Being, called nature, to be, as his vicegerent, continually watchful for the good of the universe in general, and of the particular bodies, that compose it; whilst in the mean time, this Being appears not to have the skill, or the power, to prevent such anomalies which oftentimes prove destructive to multitudes of animals, and other noble creatures. . . .

I shall add, that the doctrine I plead for does much better, than its rival, comply with what religion teaches us about the extraordinary and supernatural interpositions of divine providence. For when it pleases God to over-rule, or control, the established course of things in the world, by his own omnipotent hand, what is thus performed may be much easier discerned and acknowledged to be miraculous by them that admit in the ordinary concourse of corporeal things, nothing but matter and motion, whose powers men may well judge of, than by those who think there is besides a certain semi-deity, which they call nature, whose skill and knowledge they acknowledge to be exceedingly great. . . .

— 4 —

John Ray, *The Wisdom of God Manifested in the Works of Creation*, 3rd ed. (London: 1705), preface, 221–226

The Wisdom of God Manifested in the Works of Creation, which first appeared in 1691, was probably the most widely read seventeenth-century work in natural theology. Unlike some of his Latitudinarian colleagues, Ray thought that one could achieve certain knowledge of God's existence through study of the natural world. The following selections are from the preface and Book I. The treatment of mountains here demonstrates the combination of utilitarian and aesthetic arguments that characterized Ray's work and set a pattern for later natural theologians to follow.

For this discourse, I have been careful to admit nothing but Matter of Fact, or Experiment, but what is undoubtedly true, lest I should build upon a sandy and ruinous Foundation; and by the admixture of what is false, render that which is true suspicious. . . .

I shall now add a Word or two concerning the Usefulness of the Argument or Matter of this Discourse, and the Reason I had to make choice of it. . . . *First*, The belief of a Deity being the Foundation of all Religion; (Religion being nothing but a devout worshiping of God, or an inclination of Mind to serve and worship him;) *For he that cometh to God must believe that he is*: It is a matter of the highest Concernment, to be firmly settled and established in a full Persuasion of this main Point: Now this must be demonstrated by Arguments drawn from the Light of Nature, and works of the Creation. For as all other Sciences, so Divinity proves not, but supposes its Subject, taking it for granted that by Natural Light, Men are sufficiently convinced of the Being of a Deity. There are indeed supernatural Demonstrations of this Fundamental Truth, but not common to all Persons or Times, and so liable to Cavil and Exception by Atheistical Persons, as Inward Illuminations of Mind, a Spirit of Prophecy and Fore-telling future Contingents, illustrious Miracles, and the like. But these Proofs taken from Effects and Operations, exposed to every man's view, not to be denied or questioned by any, are most effectual to convince all that deny or doubt of it. Neither are they only convictive of the greatest and Subtlest Adversaries, but intelligible also to the meanest Capacities. For you may hear illiterate Persons of the lowest Rank of the Commonality affirming, That they need no Proof of the Being of God, for that every Pile of Grass, or Ear of Corn, sufficiently proves that: For, say they, all the Men of the World cannot make such a thing as one of these, and if they cannot do it, who can, or did, make it but God? To tell them, that it made it self, or sprung up by chance, would be as ridiculous as to tell the greatest Philosopher so.

Secondly, The Particulars of this Discourse, serve not only to demonstrate the Being of a Deity, but also to demonstrate some of his principal Attributes; as namely, his infinite Power and Wisdom. The vast multitude of Creatures, and those not only small, but immensly great: the Sun and Moon, and all the Heavenly Host, are Effects and Proofs of his Almighty Power. *The Heavens declare the Glory of God, and the Firmament sheweth his Handy-work, Psal. 19. 1.* . . . Lastly, They serve to stir up and increase in us the Affections and Habits of Admiration, Humility, and Gratitude. . . .

Because Mountains have been lookt upon by some as Warts and superfluous Excrescencies, of no use or benefit; nay, rather as signs and proofs, that the present Earth is nothing but a heap of Rubbish and Ruins, I shall deduce and demonstrate in particulars, the great Use, Benefit, and Necessity of them.

I. They are of eminent Use for the production and Original of Springs and Rivers. Without Hills and Mountains there could be no such things, or at least but very few. No more than we now find in plain and level Countries; that is, so few that it was never my hap to see one. In winter-time, indeed we might have Torrents and Landfloods, and perhaps sometimes great Inundations, but in Summer nothing but stagnating Water, reserved in Pools and Cisterns or drawn up out of deep Wells. But as for a great part of the Earth, (all lying within or near the Tropics) it would neither have Rivers nor any Rain at all. We should consequently lose all those Conveniences and Advantages that Rivers afford us of Fishing, Navigation, Carriage, Driving our Mills, Engines, and many others. The end of Mountains, I find assigned by Mr. *Edmund Halley*, a Man of great sagacity and deep insight into the Natures and Causes of Things, in a Discourse of his published in the *Philososoph. Transactions,* Numb. 192, in these words: *This, if we may allow first causes* [*Hardiment*, the thing is clear, pronounce boldly without any *ifs* or *Ands*] *This seems to be one design of the Hills, that their Ridges being placed through the middles of the Continents, might serve as it were Alembicks, to distil fresh Water for the use of Man and Beast; and their heights to give a decent to those streams, to run gently like so many Veins of the Macrocosm, to be the more benefital to the Creation.*

II. They are of great use for the Generation and convenient digging up of Metals and Minerals: which how necessary Instruments they are of Culture and Civility I have before shewn. These we see, are all digged out of Mountains, and I doubt whether there is, or can be, any Generation of them in perfectly plain and level Countries. But if there be, yet could not such Mines, without great pains and charges, if at all, be wrought; the Delfs would be soon slown with Waters, (it being impossible to make any *Addits* or *Soughs* to drain them) that no Gins or Machines could suffice to lay and keep them dry.

III. They are useful to Mankind in affording them convenient Places for Habitation, and Situations of Houses and Villages, serving as Skreens to keep off the cold and nipping blasts of the *Northern* and *Easterly* winds, and reflecting the benign and cherishing Sun beams, and so rendering their habitations both more comfortable, and more chearly in Winter; and promoting the Growth of Herbs and Fruit Trees, and the maturation of their Fruits in Summer. Besides casting off the Waters, they lay the Gardens, Yards, and Avenues to the Houses dry and clean; and so as well more salutary as more elegant. Whereas Houses built in Plains, unless shaded with Trees, lie bleak and exposed to Wind and Weather; and all Winter are apt to be grievously annoyed with Mire and Dirt.

IV. They are very Ornamental to the Earth, affording pleasant and delightful Prospects, both, I. To them that look downwards from them, upon the subjacent Countries; as they must needs acknowledge, who have but on the Downs of *Sussex*, and enjoyed that ravishing Prospect of the Sea on one hand, and the Country far and wide on the other. And 2. To those that look upwards and behold them from the Plains and low Grounds; which what a refreshing and pleasure it is to the Eye, they are best able to judge who have lived in the Isle of *Ely*, or other level Countries, extending on all sides further than one can ken; or have been far out to Sea, where they can see nothing but Sky and Water. That the Mountains are pleasant Objects to behold appears, in that the very Images of them, their Draughts and Landskips are so much esteemed.

V. They serve for the production of great variety of Herbs and Trees. For it is a true Observation, That Mountains do especially abound with different Species of Vegetables, because of the great diversity of Soyls that are found there, every *Vertex*, or Eminency, almost affording new kinds. Now these Plants serve partly for the Food and Sustenance of such Animals as are proper to the Mountains, partly for Medicinal Uses; the chief Physick Herbs and Roots, and the best in their kinds growing there: it being Remarkable, that the greatest and most luxuriant *Species* in most *Genera* of Plants are Native of the Mountains: partly also for the Exercise and Diversion of such ingenious and industrious Persons, as are delighted in searching out these Natural Rarities; and observing the outward Form, Growth, Natures, and Uses, of each *Species*, and reflecting upon the Creator of them his due Praises and Benedictions.

VI. They serve for the Harbour, Entertainment, and Maintenance of various Animals, Birds, Beasts, and Insects, that breed, feed, and frequent there. For the highest Tops and Pikes of the *Alps* themselves are not destitute of their Inhabitants, the *Ibex*, or *Stein-buck*, the *Rupicapra*, or Chamois, among Quadrupeds; the *Lagopus* among Birds; and I my self have observed beautiful *Papilio's*, and store of other Insects, upon the tops of some of the *Alpine* Mountains, Nay, the highest Ridges of many of those Mountains, serve for the maintenance of Cattel for the Service of the Inhabitants of the Valleys: The Men there, leaving their Wives and younger Children below, do not, without some difficulty, clamber up the Acclivities, dragging their Kine with them where they feed them, and milk them, or make Butter and Cheese, and do all the Daiery-work, in such sorry Hovels and Sheds as they build there to inhabit in during the Summer Months. This I myself have seen and observed in Mount *Jura*, not far from *Geneva*, which is high enough to retain Snow all Winter.

— 5 —

Thomas Burnet, *The Theory of the Earth: Containing an Account of the Original of the Earth, and All of the General Changes Which it Hath Already Undergone or is to Undergo Till the Consummation of All Things*, 2nd ed. (London: 1691), 96–100

Thomas Burnet's *The Sacred Theory of the Earth*, ("Sacred" was included in the frontispiece, but not on the title page and is usually used to distinguish Burnet's from other contemporary theories of the earth) initiated a lively tradition of scientific treatises on the origins of the earth, all in the form of commentaries on the Genesis account. Burnet's theory is characteristic of seventeenth-century Latitudinarian natural theology in adopting a very open interpretation of biblical passages, in emphasizing the hypothetical nature of scientific explanations, and in adopting a version of the mechanical philosophy. Burnet's colorful prose, evident toward the end of this selection, helped to make the work extremely popular.

CHAPTER 8

The particular History of* Noah's *Flood is explain'd in all the material parts and circumstances of it, according to the preceding Theory. Any seeming difficulties removed, and the whole Section concluded with a Discourse how far the Deluge may be lookt upon as the effect of an ordinary Providence, and how far of an extraordinary.

We have now proved our Explication of the Deluge to be more than an *Idea,* or to be a true piece of Natural History; and it may be the greatest and most remarkable that hath yet been since the beginning of the World. We have shown it to be the real account of *Noah's* Flood, according to Authority both Divine and Humane; and I would willingly proceed one step further, and declare my thoughts concerning the manner and order wherein *Noah's* Flood came to pass; in what method all those things happen'd and succeeded one another that make up the history of it, as causes or effects, or other parts or circumstances: As how the Ark was born upon the waters, what effect the Rains had, at what time the Earth broke, and the Abysse was open'd; and what the condition of the Earth was upon the ending of the Flood, and such like. But I desire to propose my thoughts concerning these things only as conjectures, which I will ground as near as I can upon Scripture and Reason, and very willing they should be rectifi'd where they happen to be amiss. I know how subject we are to mistakes in these great and remote things, when we descend to particulars; but I am willing to expose the Theory to a full trial, and to shew the way for any to examine it, provided they do it with equity and sincerity. I have no other design than to contribute my endeavours to find the truth in a subject of so great importance, and wherein the World hath hitherto so little satisfaction: And he that in an obscure argument proposeth an *Hypothesis* that reacheth from end to end, though it be not exact in every particular, 'tis not without a good effect; for it gives aim to others to take their measures better, and opens their invention in a matter which otherwise, it may be, would have been impenetrable to them. . . .

[Noah is told that there will be a great flood to punish men for their wickedness and that he should build a great Ark.]

But when the appointed time was come, the Heavens began to melt, and the Rains to fall, and these were the first surprizing causes and preparatives to the Deluge; They fell, we suppose, throughout the face of the whole Earth; which could not but have a considerable effect on that Earth, being even and smooth, without Hills and eminences, and might lay it all under water to some depth; so as the Ark, if it could not float upon those Rain-waters, at least taking advantage of a River, or of a Dock or Coistern made to receive them, it might be afloat before the Abysse was broken open. For I do not suppose the Abysse broken open before any rain fell; And when the opening of the Abysse and of the Flood-gates of Heaven are mention'd together, I am apt to think that those Flood-gates were distinct from the common rain, and were something more violent and impetuous. So that there might be preparatory Rains before the disruption of the Abysse: and I do not know but those Rains, so covering up and enclosing the Earth on every side, might providentially contribute to the disruption of it; not only by softning and weakening the Arch of the Earth in the bottom of those cracks and Chasms which were made by the Sun, and which the Rain would first run into, but especially by stopping on a sudden all the pores of the Earth, and all evaporation, which would make the Vapours within struggle more violently, as we get a Fever by a Cold; and it may be in the struggle, the Doors and Bars were broke, and the great Abysse gusht out, as out of a womb.

However, when the rains were faln, we may suppose the face of the Earth cover'd over with water; and whether it was these waters that *S. Peter* refers to, or that of the Abysse afterwards, I cannot tell when he saith in his first Episle, *Chap. 3.20. Noah* and his Family *were sav'd by water*; so as the water which destroyed the rest of the World, was an instrument of their conservation, in as much as it bore up the Ark, and kept it from that impetuous shock, which it would have had, if either it had stood upon dry land when the Earth fell, or if the Earth had been dissolv'd without any water on it or under it. However, things being thus prepar'd, let us suppose the great frame of the exterior Earth to have broke at this time, or the Fountains of the great Abysse, as *Moses* saith, to have been then open'd, from thence would issue upon the fall of the Earth, with an unspeakable violence, such a Flood of waters as would over-run and overwhelm for a time all those fragments which the Earth broke into, and bury in one common Grave all Mankind, and all the inhabitants of the Earth. Besides, if the *Flood-gates* of Heaven were any thing distinct from the Forty days Rain, their effusion, 'tis likely, was at this same time when the Abysse was broken open; for the sinking of the Earth would make an extraordinary convulsion of the Regions of the Air, and that crack and noise that must be in the falling World, and in the colliusion of the Earth and the Abysse, would make a great and universal Concussion above, which things together, must needs so shake, or so squeeze the Atmosphere, as to bring down all the remaining Vapours; but the force of these motions not being equal throughout the whole Air, but drawing or pressing more in some places than in other, where the Center of the convulsion was, there would be the chiefest collection, and there would fall, not showers of Rain, or single drops, but great spouts or cascades of water; and this is that which *Moses* seems to call, not improperly, the *Cataracts* of Heaven, or the *Windows of Heaven being set open*.

Thus the Flood came to its height; and 'tis not easie to represent to ourselves this strange Scene of things, when the Deluge was in its fury and extremity; when the Earth was broken and swallow'd up in the Abysse, whose raging waters rise higher than the Mountains, and fill'd the Air with broken waves, with an universal mist, and with thick darkness, so as Nature seem'd to be in a second Chaos; and upon this Chaos rid the distrest Ark, that bore the small remains of Mankind. No Sea was ever so tumultuous as this, nor is there anything in present Nature to be compar'd with the disorder of these waters; All the Hyperboles that are us'd in the descriptions of Storms and raging Seas, were literally true in this, if not beneath it. The Ark was really carri'd to the tops of the highest Mountains, and into the places of the Clouds, and thrown down again into the deepest gulfs; and to this very state of the Deluge and of the Ark, which was a Type of the Church in this World, *David* seems to have alluded in the name of the Church, *Psal. 42. 7. Abysse calls to Abysse at the noise of thy Cataracts or water-spouts; all thy waves and billows have gone over me.* It was no doubt an extraordinary and miraculous Providence, that could make a vessel, so ill man'd, live upon such a Sea; that kept it from being dasht against the Hills, or overwhelmed in the Deeps. That Abysse which had devour'd and swallow'd up whole Forests of Woods, Cities, and Provinces, nay the whole Earth, when it had conquer'd all, and triumph'd over all, could not destroy this single Ship. I remember the story of the *Argonaticks*, when *Jason* set out to fetch the Golden Fleece, the Poet saith, all the Gods that day look'd down from Heaven, to view the Ship; and the *Nymphs* stood upon the Mountain-tops to see the noble Youth of *Thessaly* pulling at the Oars; We may with more reason suppose the good Angels to have lookt down upon this Ship of *Noah's*; and that not out of curiosity, as idle spectators, but with a passionate concern for its safety and deliverance. A Ship whose *Cargo* was no less than a whole World; that carri'd the fortunes and hopes of all posterity, and if this had perisht, the Earth, for anything we know, had been nothing but a Desert, a great ruine, a dead heap of rubbish, from the Deluge to the Conflagration. But Death and Hell, the Grave, and Destruction have their bounds. We may entertain ourselves with the consideration of the face of the Deluge, and of the broken and drown'd Earth, in this Scheme, with the floating Ark, and the guardian Angels.

— 6 —

David Hume, *The Natural History of Religion* (1757), 29–32, 41–42

Though Hume's *Natural History of Religion* drew heavily from ancient atomist sources, it ushered in a continuous tradition of anthropology and philosophy of religion that still persists. In the following selections, he emphasizes the emotional origins of religion and the transition from polytheism to monotheism.

There is an universal tendency among mankind to conceive all beings like themselves, and to transfer to every object, those qualities, with which they are familiarly acquainted, and of which they are intimately conscious. We find human faces in the moon, armies in the clouds; and, by a natural propensity, if not corrected by experience and reflection, ascribe malice or good will to every thing, that hurts or pleases us. Hence the frequency and beauty of the *prosopopoeia* in poetry; where trees, mountains, and streams are personified, and the intimate parts of nature acquire sentiment and passion. And though these poetical figures and expressions gain not on the belief, they may serve, at least, to preserve a certain tendency in the imagination, without which they could neither be beautiful nor natural. Nor is a river-god or hamadryad always taken for a mere poetical or imaginary personage; but may sometimes enter into the real creed of the ignorant vulgar; while each grove or field is represented as possessed of a particular *genius* or invisible power, which inhabits and protects it. Nay, Philosophers cannot entirely exempt themselves from this natural frailty: but have oft ascribed it to inanimate matter the horror of a *vacuum*, sympathies, antipathies, and other affections of human nature. The absurdity is not less, while we cast our eyes upwards, and transferring, as is too usual, human passions and infirmities to the deity, represent him as jealous and revengeful, capricious and partial, and, in short, a wicked and foolish man, in every respect but his superior power and authority. No wonder, then, that mankind, being placed in such an absolute ignorance of causes, and being at the same time so anxious concerning their future fortune, should immediately acknowledge a dependence on invisible powers, posessed of sentiment and intelligence. The *unknown causes* which continually employ their thought, appearing always in the same aspect, are all apprehended to be of the same kind or species. Nor is it long before we ascribe to them thought and reason and passion, and sometimes even the limbs and figures of men, in order to bring them nearer a resemblance to ourselves.

In proportion as any man's course of life is governed by accident, we always find, that he encreases in superstition; as may particularly be observed of gamesters and sailors, who, though of all mankind, the least capable of serious reflection, abound most in frivolous and superstitious apprehensions. The gods, says CORIOLANUS in DIONYSIUS, have an influence in every affair; but above all in war; where the event is so uncertain. All human life, especially before the institution of order and good government, being subject to fortuitous accidents; it is natyural, that superstition should prevail everywhere in barbarous ages, and put men on the most ernest enquiry concerning those invisible powers, who dispose of their happiness or misery. Ignorant of astronomy and the anatomy of plants and animals, and too little curious to observe the admirable adjustment of final causes; they remain still unacquainted with a first and supreme creator, and with that infinitely perfect spirit, who alone, by his almighty will, bestowed order on the whole frame of nature. Such a magnificent idea is too big for their narrow conceptions, which can neither observe the beauty of the work, nor comprehend the grandeur of its author. They suppose their deities, however potent and invisible, to be nothing but a species of human creatures, perhaps raised from among mankind, and retaining all human passions and appetites, together with corporeal limbs and organs. Such limited beings, though masters of human fate, being, each of them, incapable of extending his influence every where, must be vastly multiplied, in order to answer that variety of events, which happen over the whole of nature. Thus every place is stored with a crowd of local deities; and thus polytheism has prevailed, and still prevails, among the greatest part of uninstructed mankind.

Any of the human affections may lead us into the notion of invisible, intelligent power; hope as well as fear, gratitude as well as affliction: but if we examine our own hearts, or observe what passes around us, we shall find, that men are much oftener thrown on their knees by the meloncholy than by the agreeable passions. Prosperity is easily received as our due, and few questions are asked concerning its cause or author. It begets cheerfulness and activity and alacrity and a lively enjoyment of every social and sensual pleasure: And during this state of mind, men have little leisure or inclination to think of the unknown invisible regions. On the other hand, every disastrous accident alarms us, and sets us on enquiries concerning the principles whence it arose: Apprehensions spring up with regard to futurity: And the mind, sunk into diffidence, terror, and melancholy, has recourse to every method of appeasing those secret intelligent powers, on whom our fortune is supposed entirely to depend. . . .

The doctrine of one supreme deity, the author of nature, is very ancient, has spread itself over great and populous nations, and among them has been embraced by all ranks and conditions of men: But whoever thinks that it has owed its success to the prevalent force of those invincible reasons, on which it is undoubtedly founded, would show himself little acquainted with the ignorance and stupidity of the people, and their incurable prejudices in favor of their particular superstitions. Even to this day, and in Europe, ask any of the vulgar, why he believes in an omnipotent creator of the world; he will never mention the beauty of final causes, of which he is wholly ignorant: He will not hold out his hand, and bid you contemplate the suppleness and variety of joints in his fingers, their bending all one way, the counterpoise which they receive from the thumb, the softness and fleshy parts of the inside of the hand, with all the other circumstances which render that member fit for the use, to which it was destined. To these he has been long accustomed; and he beholds them with listlessness and unconcern. He will tell you of the sudden and unexpected death of such a one: The fall and bruise of such another: The excessive drought of this season: The cold and rains of another. These he ascribes to the immediate operation of providence: And such events, as, with good reasoners, are the chief difficulties in admitting a supreme intelligence, are with him the sole arguments for it.

Many theists, even the most zealous and refined, have denied a *particular* providence, and have asserted, that the Sovereign mind or first principle of all things, having fixed general laws, by which nature is governed, gives free and uninterrupted course to these laws, and disturbs not, at every turn, the settled order of events by particular volitions. From the beautiful connexion, say they, and rigid observance of established rules, we draw the chief argument for theism; and from the same principles are enabled to answer the principle objections against it. But so little is this understood by the generality of mankind, that, wherever they observe any one to ascribe all events to natural causes, and to remove the particular interposition of a deity, they are apt to suspect him of the greatest infidelity. *A little philosophy*, says lord BACON, *makes men atheists: a great deal reconciles them to religion*. For men, being taught, by superstitious prejudices, to lay the stress on a wrong place: when that fails them, and they discover, by a little reflection, that the course of nature is regular and uniform, their whole faith totters and falls to ruin. But being taught, by more reflection, that this very regularity and uniformity is the strongest proof of design and of a supreme intelligence, they return to that belief, which they had deserted; and they are now able to establish it on a firmer and more durable foundation.

Convulsions in nature, disorders, prodigies, miracles, though the most opposite to the plan of a wise superintendent, impress mankind with the strongest sentiments of religion; the causes of events seeming then the most unknown and unaccountable. Madness, fury, rage, and an inflamed imagination, though they sink man nearest to the level of beasts, are, for a like reason, often supposed to be the only dispositions, in which we can have any immediate communication with the Deity.

We may conclude, therefore, upon the whole, that, since the vulgar, in nations, which have embraced the doctrine of theism, still build it upon irrational and superstitious principles, they are never led into that opinion by any process of argument, but by a certain train of thinking, more suitable to their genius and capacity.

It may readily happen, in an idolatrous nation, that though men admit the existence of several limited deities, yet is there some one God, whom, in a particular manner, they make the object of their worship and adoration. They may either suppose, that, in the distribution of power and territory among the gods, their nation was subjected to the jurisdiction of that particular deity; or reducing heavenly objects to the model of things below, they may represent one god as the prince or supreme magistrate of the rest, who, though of the same nature, rules them with an authority, like an earthly sovereign exercises over his subjects and vassals. Whether this god, therefore, be considered as their particular patron, or as the general sovereign of heaven, his votaries will endeavour, by every art, to insinuate themselves into his favour; and supposing him to be pleased, like themselves, with praise and flattery, there is no eulogy or exaggeration, which will be spared in their addresses to him. In proportion as men's fears or distresses become more urgent, they still invent new strains of adulation; and even he who outdoes his predecessor in swelling up the titles of his divinity, is sure to be outdone by his successor in newer and more pompous epithets of praise. Thus they proceed; till at last they arrive at infinity itself, beyond which there is no further progress: And it is well if, in striving to get farther, and to represent a magnificent simplicity, they run into inexplicable mystery, and destroy the intelligent nature of their deity, on which alone any rational worship or adoration can be founded. While they confine themselves to the notion of a perfect being, the creator of the world, they coincide, by chance, with the principles of reason and true philosophy; though they are guided to that notion, not by reason, of which they are in a great measure incapable, but by the adulation and fears of the most vulgar superstition.

— 7 —

Ludwig Feuerbach, *The Essence of Christianity*, translated from German by George Eliot (London: 1854 [original 1841]), 12–14

Ludwig Feuerbach's *Essence of Christianity* used Hegelian language to express a completely materialist doctrine in which God was to be understood as the projection of mankind's ideal self onto the universe. The section below offers the first expression of this idea and attacks those who argue that God is somehow beyond the human ability to know.

THE ESSENCE OF RELIGION CONSIDERED GENERALLY . . .

In the perceptions of the senses consciousness of the object is distinguishable from consciousness of the self; but in religion, consciousness of the object and self-consciousness coincide. The object of the senses is out of man, the religious object is within him, and therefore as little forsakes him as his self-consciousness or his conscience; it is the intimate, the closest object. "God" says Augustine, for example, "is nearer, more related to us, and therefore more easily known to us, than sensible, corporeal things." The object of the senses is in itself indifferent—independent of the disposition or of the judgement; but the object of religion is a selected object; the most excellent, the first, the supreme being; it essentially presupposes a critical judgement, a discrimination between the divine and the non-divine, between that which is worthy of adoration and that which is not worthy. And here may be applied, without any limitation, the proposition: the object of any subject is nothing else than the subjects own nature taken objectively. Such as are a man's thoughts and dispositions, such is his God. Consciousness of God is self-consciousness, knowledge of God is self-knowledge. By his God thou knowest the man, and by the man his God; the two are identical. Whatever is God to a man, that is his heart and soul; and conversely, God is the manifested inward nature, the expressed self of a man,—religion the solemn unveiling of a man's hidden treasures, the revelation of his intimate thoughts, the open confession of his love-secrets.

But when religion—consciousness of God—is designated as the self-consciousness of man, this is not to be understood as affirming that the religious man is directly aware of this identity; for, on the contrary, ignorance of it is fundamental to the peculiar nature of religion. To preclude this misconception, it is better to say, religion is man's earliest and also indirect form of self-knowledge. Hence, religion everywhere precedes philosophy, as in the history of the race, so also in that of the individual. Man first of all sees his nature as if *out of* himself, before he finds it in himself. His own nature is in the first instance contemplated by him as that of another being. Religion is the child-like condition of humanity; but the child sees his nature—man—out of himself; in childhood a man is an object to himself, under the form of another man. Hence the historical progress of religion consists in this: what by an earlier religion was regarded as objective, is now regarded as subjective; that is, what was formerly worshiped and contemplated as God is now perceived to be something *human*. What was at first religion, becomes, at a later period, idolatry; man is seen to have adorned his own nature. Man has given objectivity to himself, but has not recognized the object as his own nature: a later religion takes this forward step; every advance in religion is therefore a deeper self-knowledge. But every particular religion, while it pronounces its predecessors idolatrous, excepts itself—and necessarily so, otherwise it would no longer be religion—from the fate, the common nature of all religions: it imputes only to other religions what is the fault, if fault it be, of religion in general. Because it has a different object, a different tenor, because it has transcended the ideas of preceding religions, it erroneously supposes itself exalted above the necessary eternal laws which constitute the essence of religion—it fancies its object, its ideas, to be superhuman. But the essence of religion, thus hidden from the religious, is evident to the thinker, by whom religion is viewed objectively, which it cannot be by its votaries. And it is our task to show that the antithesis of divine and human is altogether illusory, that it is nothing else than the antithesis between the human nature in general and the human individual; that, consequently, the object and contents of the Christian religion are altogether human.

Religion, at least the Christian, is the relation of man to himself, or more correctly to his own nature (*i. e.*, his subjective nature); but a relation to it, viewed as a nature apart from his own. The divine being is nothing else than the human being, or rather the human nature purified, freed from the limits of the individual man, made objective—*i. e.*, contemplated and revered as another, a distinct being. All the attributes of the divine nature are, therefore, attributes of the human nature.

In relation to the attributes, the predicates, of the Divine Being, this is admitted without hesitation, but by no means in relation to the subject of these predicates. The negation of the subject is held to be irreligion, any, atheism; though not so the negation of the predicates. But that which has no predicates or qualities, has no effect upon me; that which has no effect on me has no existence for me. To deny all the qualities of a being is equivalent to denying the being himself. A being without qualities is one which cannot become an object to the mind, and such a being is virtually non-existent. Where man deprives God of all qualities, God is no longer anything more to him than a negative being. To the truly religious man, God is not a being without qualities, because to him, he is a positive, real being. The theory that God cannot be defined, and consequently cannot be known by man, is therefore the offspring of recent times, a product of modern unbelief. . . . The denial of determinate, positive predicates concerning the divine nature is nothing else than a denial of religion, with, however, an appearance of religion in its favour, so that it is not recognized as a denial; it is simply a subtle, disguised atheism. . . . He who earnestly believes in the Divine existence is no shocked at the attributing even of gross sensuous qualities to God. He who dreads an existence that may give offence, who shrinks from the grossness of a positive predicate, may as well renounce existence altogether.

— 8 —

John William Draper, *History of the Conflict between Religion and Science* (New York: D. Appleton and Company, 1896 [original 1874]), vi–viii, x–xi, xv, 167–168, 170–172

An American Protestant chemist and intellectual historian, Draper was the first author to promote the phrase "conflict between religion and science." His work also illustrates the strong anti-Catholic strain in American intellectual life in the late nineteenth century, and his treatment of Galileo demonstrates the casual attitude toward chronology and facts that pervaded the works of nineteenth-century conflict theorists.

The history of science is not a mere record of isolated discoveries; it is a narrative of the conflict of two contending powers, the expansive force of the human intellect on one side, and the compression arising from traditional faith and human interests on the other. . . .

A few years ago, it was politic and therefore the proper course to abstain from all allusion to this controversy, and to keep it as far as possible in the background. The tranquility of society depends so much on the stability of its religious convictions, that no one can be justified in wantonly disturbing them. But faith is in its nature unchangeable, stationary; Science is in its nature progressive; and eventually a divergence between them, impossible to conceal, must take place. It then becomes the duty of those whose lives have made them familiar with both modes of thought, to present modestly, but firmly, their views; to compare the antagonistic pretensions calmly, impartially, philosophically. History shows that, if this be not done, social misfortunes, disastrous and enduring, will ensue. When the old mythical religion of Europe broke down under the weight of its own inconsistencies, neither the Roman emperors nor the philosophers of those times did anything adequate for the guidance of public opinion. They left religious affairs to take their chance, and accordingly those affairs fell into the hands of ignorant and infuriated ecclesiastics, parasites, eunuchs, and slaves. . . .

In speaking of Christianity, reference is generally made to the Roman Church, partly because its adherents compose the majority of Christendom, partly because its demands are the most pretentious, and partly because it has commonly sought to enforce those demands by the civil power. None of the Protestant Churches has ever occupied a position so imperious—none has ever had such wide spread political influence. For the most part they have been averse to constraint, and except in very few instances their opposition has not passed beyond the exciting of theological odium.

As to Science, she has never sought to ally herself to civil power. She has never attempted to throw odium or inflict social ruin on any human being. She has never subjected anyone to mental torment, physical torture, least of all to death, for the purposes of upholding or promoting her ideas. She presents herself unstained by cruelties and crimes. But the Vatican—we have only to recall the Inquisition—the hands that are now raised in appeals to the Most Merciful are crimsoned. They have been steeped in blood! . . .

In the sixteenth century . . . it was clearly seen by many pious men that Religion was not accountable for the false position in which she was found, but that the misfortune was directly traceable to the alliance she had of old contracted with Roman paganism. The obvious remedy, therefore, was a return to primitive purity. Thus arose the . . . conflict, known to us as the Reformation. . . . The special form it assumed was a contest respecting the standard or criterion of truth, whether it is to be found in the Church or in the Bible. The determination of this involved a settlement of the rights of reason, or intellectual freedom. Luther, who is the conspicuous man of the epoch, carried into effect his intention with no inconsiderable success; and at the close of the struggle it was found that Northern Europe was lost to Roman Christianity. . . .

Copernicus, a Prussian, about the year 1507, had completed a book "On the Revolutions of the Heavenly Bodies." He had journeyed to Italy in his youth, had devoted his attention to astronomy, and had taught mathematics at Rome. From a profound study of the Ptolemaic and Pythagorean systems, he had come to a conclusion in favor of the latter, the object of his book being to sustain it. Aware that his doctrines were totally opposed to the revealed truth, and forseeing that they would bring upon him the punishments of the Church, he expressed himself in a cautious and apologetic manner, saying that he had only taken the liberty of trying whether, on the supposition of the earth's motion, it was possible to find better explanations than the ancient ones of the revolutions of the celestial orbs; that in doing this he had only taken the privilege that had been allowed to others, of feigning what hypothesis they chose. The preface was addressed to Pope Paul III.

Full of misgivings as to what might be the result, he refrained from publishing the book for thirty-six years, thinking that "perhaps it might be better to follow the examples of the Pythagoreans and others, who delivered their doctrine only by tradition and to friends." At the entreaty of Cardinal Schomberg he at length published it in 1543. A copy of it was brought to him on his death bed. Its fate was such as he had anticipated. The inquisition condemned it as heretical. In their decree, prohibiting it, the Congregation of the Index denounced his system as "that false Pythagorean doctrine utterly contrary to the Holy Scriptures."

It had been objected to the Copernican theory that, if the planets Mercury and Venus move round the sun in orbits interior to that of the earth, they ought to show phases like those of the moon; and that in the case of Venus, which is so brilliant and conspicuous, these phases should be very obvious. Copernicus himself had admitted the force of the objection, and had vainly tried to find an explanation. Galileo, on turning his telescope to the planet, discovered that the expected phases actually exist; now she was a crescent, then half-moon, then gibbous, then full. Previously to Copernicus, it was supposed that the planets shine with their own light, but the phases of Venus and Mars [*sic*] prove that their light is reflected. The Aristotelian notion, that the celestial differ from terrestrial bodies in being incorruptible, received a rude shock from the discoveries of Galileo, that there were mountains and valleys in the moon like those of the earth, that the sun is not perfect, but has spots on its face, and that he turns on his axis instead of being in a state of majestic rest. The apparition of new stars had already thrown serious doubts on this theory of incorruptibility.

These and many other beautiful telescopic discoveries tended to the establishment of the truth of the Copernican theory and gave unbounded alarm to the Church. By the low and ignorant ecclesiastics they were denounced as deceptions or frauds. Some affirmed that the telescope might be relied on well enough for terrestrial objects, but with the heavenly bodies it was altogether a different affair. Others declared that its invention was a mere application of Aristotle's remark that stars could be seen in the daytime from the bottom of a deep well. Galileo was accused of imposture, heresy, blasphemy, atheism. With a view of defending himself, he addressed a letter to the Abbe Castelli, suggesting that the Scriptures were never intended to be a scientific authority, but only a moral guide. This made matters worse. He was summoned before the Holy Inquisition, under an accusation of having taught that the earth moves round the sun, a doctrine "utterly contrary to the Scriptures." He was ordered to renounce that heresy, on pain of being imprisoned. He was directed to desist from teaching and advocating the Copernican theory, and pledge himself that he would neither publish nor defend it for the future. Knowing well that Truth has no need of martyrs, he assented to the required recantation, and gave the promise demanded.

For sixteen years the Church had rest. But in 1632 Galileo ventured on the publication of his work entitled "The System of the World," its object being the vindication of the Copernican doctrine. He was again summoned before the Inquisition at Rome, accused of having asserted that the earth moves round the sun. He was declared to have brought upon himself the penalties of heresy. On his knees, with his hand on the Bible, he was compelled to abjure and curse the doctrine of the movement of the earth. What a spectacle! This venerable man, the most illustrious of his age, forced by the threat of death to deny facts which his judges as well as himself knew to be true! He was then committed to prison, treated with remorseless severity during the remaining ten years of his life, and was denied burial in consecrated ground. Must not that be false which requires for its support so much imposture, so much barbarity? The opinions thus defended by the Inquisition are now objects of derision to the whole civilized World.

— 9 —

James McCosh, *The Religious Aspect of Evolution*, enlarged and improved edition (New York: Charles Scribner's Sons, 1890), 5–8, 27, 58–61

Among nineteenth-century American moderate evangelical Christians, James McCosh, moral philosopher and president of Princeton University, was one of the most outspoken supporters of evolution. In the following selections, he explains why he believes that theistic evolution is perfectly consistent with divine design and action, disagreeing with the Princeton theologian Charles Hodge.

THE QUESTION BETWEEN EVOLUTIONISTS AND NON-EVOLUTIONISTS.—"No man can find out the work that God maketh from the beginning to the end." But though human science cannot go back to the beginning nor go on to the end, and while there is much in the middle that is concealed, there are whole provinces which we can inquire into and come to know. "We know in part." We now know not a little about the generation of our earth, and of the plants and animals upon its surface. And we can tell much about the order in which animated beings appeared. But there is a keen dispute as to how they were produced.

All admit that there is system in the production of the organic world. Those who have no faith in a power above nature, ascribe it to physical forces. Religious people, so far from denying this, should at once admit and proclaim it; and seek to find out what the forces are and the laws they follow. We cannot allow God to be separated from his works, and so we must resolutely hold that God is in the forces arranged into an order—that is, laws, which we find it so interesting to observe.

But this is not just the burning question of the day. There is a perplexing confusion in the statement of the question. It has been misunderstood by religious, it has been perverted by irreligious, people. The former often speak of it as being: Whether all things are to be ascribed to God, or a portion to God, while the rest is handed over to material agency? In maintaining this latter view they furnish an excuse or pretext to those who would ascribe the descent of plants and animals to mechanical agency. The great body of naturalists, all younger than forty, certainly all younger than thirty, are sure that they see evolution in nature; but they are assured by their teachers or the religious press that if evolution does every thing, there is nothing left for God to do, and they see no proof of his existence. Many a youth is brought to a crisis in his belief and life by such a representation. He feels that he must give up either his science or his faith, and his head is distracted, and his heart is tortured till feelings more bitter than tears are wrung from it.

The question is said to be, Whether the origin of species and descent of living creatures are by supernatural power or natural law, by Creator or creative action, by design or by mechanism, by contrivance or by chance, by purpose or without purpose.

Mr. Darwin, followed by Dr. Romanes, and many others, is constantly drawing the distinction in this form: between "natural selection" and "special creation." Now the difference between the two opposing theories as thus put is misleading, and this whether put by disbelief or by belief. The supernatural power is to be recognized in the natural law. The Creator's power is executed by creature action. The design is seen in the mechanism. Chance is obliged to vanish because we see contrivance. There is purpose when we see a beneficent end accomplished. Supernatural design produces natural selection. Special creation is included in universal creation.

A question is often settled by being properly stated. The *status quaestionis*, as the scholastics expressed it, is here not between God and not-God, but between God working without means and by means, the means being created by God and working for him. There may be evidence of design, of contrivance, and purpose in the very means employed. If an optician brings me a microscope I have only to examine it to discover design in it, but I may have as clear proof of purpose when I visit his shop and see him manufacturing the instrument. There is nothing atheistic in the creed that God proceeds by instruments, which we may find to be for the good of his creatures. There may be a want of reverence toward God and truth when there is evidence laid before us in its favor and we refuse to look at it. I should discover God in the human frame, on the supposition that he created it at once, but I have quite as satisfactory evidence on the supposition that he produced it by a father and mother, and provided that it should grow to maturity by a natural process. In the geological development I am privileged as it were to enter God's workshop and see his modes of operation, and the results reached so full of provisions in bones, muscles, joints, for the good of the creature. . . .

I have never been able to see that religion, and in particular that Scripture in which our religion is embodied, is concerned with the absolute immutability of species. Final Cause, which is a doctrine of natural religion, should be satisfied with species being so fixed as to secure the stability of nature. If new species appear in our world, they differ so slightly from the old, out of which they have been formed, that there have been no violent or revolutionary changes involved. Nature is kept steadfast and theism is satisfied, even though in rare circumstances a new species should be produced to diversify nature and make it equal to the duty of peopling the earth, which is certainly one of the purposes of God by which he widens the sphere of happiness. . . .

GOD IN EVOLUTION.—There is, or was, a wide-spread idea that the doctrine of development is adverse to religion. This has arisen mainly from the circumstance that it seems to remove God altogether, or at least to a greater distance from his works, and this has been increased by the circumstance that the theory has been turned to atheistic purposes. This impression is to be removed, first, by declaring emphatically that we are to look on evolution simply as the method by which God works. It is a forgotten circumstance that when Newton proclaimed the law of gravitation it was urged that he thereby took from God an important part of his works to hand it over to material mechanism, and the objection had to be removed in a quarto volume written by the celebrated mathematician, Maclaurin; and this was the more easily done from the circumstance that Newton was a man of profound religious convictions. The time has now come when people must judge of a supposed scientific theory, not from the faith or unbelief of the discoverer, but from the evidence in its behalf. They will find that whatever is true, is also good, and will in the end be favorable to religion.

A second erroneous impression needs to be effaced. Because God executes his purposes by agents, which it should be observed he has himself appointed, we are not therefore to argue that he does not continue to act, that he does not now act. He may have set agoing the evolution millions of years ago, but he did not then cease from his operation, and sit aloof and apart to see the machine moving. He is still in his works, which not only were created by him, but have no power without his in-dwelling. Though an event may have been ordained from all eternity, God is as much concerned in it as if he only ordained it now. God acts in his works now quite as much as he did in their original creation. The effects follow, the product is evolved, because he wills it, just as plants generate only when there is light shining on them; just as day continues only because the sun shines. A birth or a death may be brought about by a caused evolution, but the mother may rest assured that God is in both, rejoicing with her, or pitying her.

I hold that time is a reality, so perceived by our minds and so perceived by the Divine Mind. The *eternal now* spoken of by some of the schoolmen and by the poet Crowley is a contradiction. But while time, past, present, and future, is a reality to Deity, it may stand in a very different relation to him from what it is to us. Time, past and future, may be contemplated as immediately by him as time present is by us, and his love be literally an everlasting love, comprehending all time, as his omniscience does all space.

FINAL CAUSE.—I do not propose in these Lectures to prove anew the existence of god. This has been done so satisfactorily by a succession of able men since the days of Socrates that it does not need to be repeated. My aim rather is to show that the doctrine of evolution does not undermine the argument from Final Cause, but rather strengthens it by furnishing new illustrations of the wisdom and goodness of God. The proof from design proceeds on the observation of things as adapted to one another to accomplish a good end, and is equally valid whether we suppose adjustments to have been made at once or produced by a process which has been going on for millions of years. There is proof of a designing mind in the eye as it is now presented to us, with its coats and humors, rods and cones, retina and nerves, all co-operating with each other and with the beams that fall upon them from suns millions of miles away. But there is further proof in the agents having been brought into relation by long processes all tending to the one end. I value a gift received from the hand of a father; but I appreciate it more when I learn that the father has been using many and varied means to earn it for me.

Annotated Bibliography

Alexander, H. G., ed. 1956. *The Leibniz-Clarke Correspondence.* Manchester: Manchester University Press. The correspondence between Leibniz and the Newtonian spokesman Samuel Clark constituted one of the most important early-eighteenth-century debates over the religious implications of Newtonian natural philosophy. Alexander's introduction provides a good description of the setting in which the correspondence took place.

Allenby, R. Scott. 1999. "Exposing Darwin's 'Hidden Agenda': Roman Catholic Responses to Evolution, 1875–1925." Pp. 173–204 in Ronald Numbers and John Stenhouse, eds., *Disseminating Darwinism: The Role of Place, Race, Religion, and Gender*. Cambridge: Cambridge University Press. Excellent article focusing on American Catholic views.

Andreae, Johann. 1916. *Christianopolis: An Ideal State of the Seventeenth Century.* Edited and translated by Felix Held. Oxford: Oxford University Press.

Appleman, Phillip, ed. 1970. *Darwin: Texts, Commentary.* New York: W. W. Norton. The best introduction to Darwin that I know. Includes excerpts from relevant background materials, substantial sections from Darwin's major writings, and a selection of critical responses, including several specifically directed at Darwinism and religious thought.

Aquinas, Thomas. 1947. *The Summa Theologica.* Translated by Fathers of the English Dominican Province. New York: Benziger Brothers (online version).

Aristophanes. 1962. *The Clouds*. Translated by William Arrowsmith. New York: New American Library.

Ashcraft, Richard. 1969. "Faith and Knowledge in Locke's Philosophy." Pp. 194–223 in John Yolton, ed., *John Locke: Problems and Perspectives*. London: Cambridge University Press. The best short introduction to the religious motives underlying Locke's philosophy that I know.

Ashworth, William. 1986. "Catholicism and Early Modern Science." Pp. 136–166 in David C. Lindberg and Ronald L. Numbers, eds., *God and Nature: Historical Essays on the Encounter Between Christianity and Science*. Berkeley and Los Angeles: University of California Press. A good introduction, but less sympathetic toward the Catholic Church than seems warranted by the more recent works of Feldhay, 1987 and 1989 and Heilbron 1999.

Astore, William. 2001. *Observing God: Thomas Dick, Evangelicalism, and Popular Science in Victorian Britain and America*. Aldershot: Ashgate Publishing Ltd. Outstanding and detailed analysis of perhaps the most widely read evangelical popularizer of science in the nineteenth century.

Bacon, Francis. 1937. *Essays, Advancement of Learning, New Atlantis, and Other Pieces*. Selected and edited by Richard Foster Jones. New York: Odyssey Press.

Basil the Great. 1963. *Exigetic Homilies*. Washington, D.C.: Catholic Press of America.

Bigg, Charles. 1968. *The Christian Platonists of Alexandria*. 1886. Reprint, Oxford: Oxford University Press.

Blackwell, Richard J. 1991. *Galileo, Bellarmine, and the Bible*. Notre Dame: University of Notre Dame Press. Excellent discussion of the intellectual position of Robert Bellarmine, the officer of the Inquisition who warned Galileo about his Copernican views in 1616.

Boas, Marie. 1962. *The Scientific Renaissance: 1450–1600*. New York: Harper and Row. Though there have been many more recent introductions to Renaissance science, this remains one of the most comprehensive and clearly written.

Bono, James. 1995. *The Word of God and the Languages of Man: Interpreting Nature in Early Modern Science and Medicine*. Vol. 1, *Ficino to Descartes*. Madison: University of Wisconsin Press. Focuses on the changing assumptions about the uses of language in early modern Europe and their implications for both science and religion. Not an easy book to read.

Boscovich, Roger Joseph. 1922. *A Theory of Natural Philosophy*. 1763. Reprint, Chicago: Open Court Press.

Boyle, Robert. 1675. *Some Considerations of the Reconcileableness of Reason and Religion*. London.

———. 1744. *The Works of the Honorable Robert Boyle*. 5 vols. London.

Bozeman, Theodore Dwight. 1977. *Protestants in an Age of Science: The Baconian Ideal and Antebellum American Religious Thought*. Chapel Hill: University of North Carolina Press. One of the two (along with Hovenkamp 1978) best and most often cited works on science and religion in America during the first half of the nineteenth century. It is more limited than Hovenkamp by virtue of focusing exclusively on "Old School Presbyterianism," a Princeton-centered conservative Calvinist movement that emphasized human imperfection but aggressively promoted education, and this focus allows the book to be somewhat deeper on the topics it covers.

Brooke, John Hedley. 1988. "The God of Isaac Newton." Pp. 169–183 in John Fauvel, Raymond Flood, Michael Shortland, and Robin Wilson, *Let Newton Be!* Oxford: Oxford University Press. If you only have time to read one short essay on Newton's religion, this should be it.

———. 1991. *Science And Religion: Some Historical Perspectives*. Cambridge: Cambridge University Press. Almost universally conceded to be the best single-volume synthetic work on science and Christianity from circa 1500 to the present. Brooke includes a fifty-five-page bibliographic essay, which is very valuable.

Brooke, John Hedley, and Geoffrey Cantor. 2000. *Reconstructing Nature: The Engagement of Science and Religion (Glasgow Gifford Lectures)*. New York: Oxford University Press. More readable but not as encyclopedic as Brooke, 1991. Continues Brooke's earlier insistence that no single or small number of kinds of interaction adequately characterize the relations between science and religion. Includes a very accessible section on Comte's Religion of Humanity.

Brooke, John Hedley, Margaret J. Osler, and Jitse M. van der Meer, eds. 2001. *Science in Theistic Contexts: Cognitive Dimensions*. Osiris, no. 16. Chicago: University of Chicago Press, for the History of Science Society. Most arguments about science and religion interactions focus on the general encouragement or discouragement that one institution provides to the other. The sixteen articles in this collection try to link the specific content of scientific knowledge to theological issues and span the period from early modern to late Victorian times.

Brooks, Richard S., and David K. Himrod. 2001. *Science and Religion in the English Speaking World, 1600–1727: A Bibliographical Guide to the Secondary Literature.* Lanham, Md.: Scarecrow Press. Start here if you want to research any topic covered. Over 1,700 items, each carefully annotated to give a sense of content and quality.

Buckley, Michael. 1987. *At the Origins of Modern Atheism.* New Haven: Yale University Press. Argues that Newtonian natural theology unwittingly set the stage for modern atheism by promoting the idea that religion needed scientific support.

Burchfield, Joe. 1975. *Lord Kelvin and the Age of the Earth.* New York: Science History Publications. Thorough analysis of one of the key technical blocks to acceptance of Darwinian evolution by natural selection. Kelvin's motives for estimating the age of the earth included religious questions regarding natural selection.

Burns, R.M. 1981. *The Great Debate on Miracles from Joseph Glanvill to David Hume.* Lewisburg, Pa.: Bucknell University Press. Argues that it was among natural philosophers such as Boyle, Glanvill, Wilkins, and Locke that the emphasis on miracles as evidence for the authority of Scripture became crucial. Within their works, the theologian and the natural philosopher were presented as engaging in similar enterprises.

Burrow, J.W. 1970. *Evolution and Society: A Study in Victorian Social Theory.* Cambridge: Cambridge University Press. Includes a good introduction to evolutionary anthropology with its special emphasis on the evolution of religion. Should be supplemented by Stocking, 1968.

Burtt, Edwin Arthur. 1925. *The Metaphysical Foundations of Modern Physical Science.* London: Paul, Trench, Teubner. Frequently reprinted treatment of the philosophical and theological context within which modern science developed. Path breaking in its time; but now very much outdated. It should not be used as a sole source on any topic.

Butler, John. 1979. "Magic, Astrology, and the Early American Religious Heritage, 1600–1760." *American Historical Review* 84:317–346. A valuable introduction to the relationships between the occult sciences and religion in America.

Calvin, John. 1995. *Institutes of the Christian Religion.* 1536. Reprint, N.p.: William B. Eerdmans Publishing Co. Translated by F.L. Battles.

Cantor, Geoffrey N. 1979. "Revelation and the Cyclical Cosmos of John Hutchinson." Pp. 3–22 in L.J. Jordanova and Roy Porter, eds., *Images of the Earth.* Chalfont St. Giles: British Society for the History of Science.

An excellent introduction to the most prominent British anti-Newtonian natural philosopher of the early eighteenth century. His biblically based science appealed to High Church figures and evangelical Methodists alike.

———. 1991. *Michael Faraday: Sandemanian and Scientist: A Study of Science and Religion in the Nineteenth Century*. London: Palgrave McMillan. Though Cantor has written other more accessible biographies of Faraday, this is the most detailed study of how the social doctrines of the Sandemanians shaped Faraday's attitudes toward science and the scientific community.

Cassirer, Ernst. 1955. *The Philosophy of the Enlightenment*. Boston: Beacon Press.

Chadwick, Owen. 1975. *The Secularization of the European Mind in the Nineteenth Century*. Cambridge: Cambridge University Press. Argues that when Victorian scientists opposed religion it was out of a variety of motives that were seldom directly related to the content of their science. Chadwick focuses especially on the way in which historical knowledge challenged traditional interpretations of the Bible in the nineteenth century. But, argues Chadwick, the public, sensitized by positivist views, failed to make such distinctions and was inclined to see every challenge to religious orthodoxy as scientific.

Chambers, Robert. 1994. *Vestiges of the Natural History of Creation and Other Evolutionary Writings*. Edited by James Secord. 1844. Reprint, Chicago: The University of Chicago Press. One of the key texts for understanding Victorian reactions to evolutionary ideas.

Charleton, Robert. 1652. *The Darkness of Atheism Dispelled by the Light of Nature: A Physico-Theological Treatise*. London.

———. 1654. *Physiologia Epicuro-Gassendo-Charletoniana: or A Fabric of Science Natural, Upon the Hypothesis of Atoms*. London.

Christianson, Gale E. 1984. *In the Presence of the Creator: Isaac Newton and His Times*. New York: Free Press. An excellent biography of Newton for the general reader who is not willing to work through the 900-plus pages of Westfall's *Never at Rest*. Christianson foregrounds Newton's religious concerns and is thin on explaining the content of Newton's science.

Clarke, Samuel. 1706. *A Discourse Concerning the Unchangeable Obligations of Natural Religion, and the Truth and Certainty of Christian Revelation*. London.

Cohen, I. Bernard, ed. 1958. *Isaac Newton's Papers and Letters On Natural Philosophy*. Cambridge, Mass.: Harvard University Press.

———. 1990. *Puritanism and the Rise of Modern Science: The Merton Thesis*. New Brunswick and London: Rutgers University Press. Collects segments from Robert Merton's groundbreaking work on Puritanism and science along with selections from critics and supporters over a sixty-year period, then offers Merton a chance to reflect on the impact of his work.

Comte, Auguste. 1966. *System of Positive Polity*. 1875–1877. Reprint, New York: Burt Franklin.

———. 1973. *The Catechism of Positive Religion*. Translated by Richard Congreve. 1891. Reprint, Clifton, N.J.: Augustus M. Kelley.

———. 1974. *The Positive Philosophy*. Translated by Harriet Martineau. 1855. Reprint, New York: AMS Press.

Cooper, Robert. 1853. *The Immortality of the Soul*. London.

Cope, Jackson I. 1956. *Joseph Glanvill: Anglican Apologist*. St. Louis: Washington University. A valuable biography of the poster boy for Latitudinarian-experimental mechanical philosophy interactions.

Copenhaver, Brian P. 1990. "Natural Magic, Hermeticism, and Occultism in Early Modern Science." Pp. 261–301 in David Lindberg and Robert S. Westman, eds., *Reappraisals of the Scientific Revolution*. Cambridge: Cambridge University Press. Bemoans the imprecision of recent uses of such terms as Hermetic, natural magic, and so forth, but admits that it reflects an imprecision among early modern practitioners. Argues that the work of Dame Frances Yates should be used with caution, but praises her for expanding the domain of the history of science.

Costa, Cruz. 1964. *A History of Ideas in Brazil: The Development of Philosophy in Brazil and the Evolution of National History*. Berkeley and Los Angeles: University of California Press. Includes a good discussion of the role of Comtean positivism and his Religion of Humanity in Brazilian intellectual and political life.

Coudert, Allison P. 1980. *Alchemy: The Philosopher's Stone*. Boulder Colo.: Shambhala. A popular history of alchemy by a very capable scholar. Ends with the seventeenth century.

Cragg, Gerald R., ed. 1968. *The Cambridge Platonists*. New York: Oxford University Press. Includes a good introduction to this group of moderate Anglican proponents of a non-mechanical natural philosophy and natural theology; then offers selected texts from the group.

Craig, John. 1964. "Craig's Rules of Historical Evidence." *History and Theory. Beiheft;* 4: A translation and reprinting of key sections of Craig's *Theologia Christiane Principia Mathematica*, a 1699 attempt to use Newton's mathematical methods to justify belief in the Gospels.

Craven, Kenneth. 1992. *Jonathan Swift and the Millenialism of Madness: The Information Age in Swift's Tale of a Tub*. Leiden: Brill. Detailed discussion of the most brilliant among conservative Anglican opponents of natural theology and natural philosophy at the beginning of the eighteenth century.

Cudworth, Ralph. 1845. *The True Intellectual System of the Universe: Wherein All the Reason and Philosophy of Atheism Is Confuted, and Its Impossibility Demonstrated, To Which Are Added the Notes and Dissertations of Dr. J. L. Mosheims*. London: Thomas Tegg.

Daniel, Stephen H. 1984. *John Toland: His Methods, Manners, and Mind*. Kingston, Canada: McGill-Queens University Press. Valuable discussion of the connections among Toland's materialist natural philosophy and his pantheistic religion, linking him to Giordano Bruno.

Darwin, Charles. n.d. *The Origin of Species by Means of Natural Selection, or the Preservation of Favored Races in the Struggle for Life, and, The Descent of Man and Selection in Relation to Sex*. New York: The Modern Library. Reissue in inexpensive format of the 1859 edition of *Origin* and the 1873 edition of *Descent*.

Dear, Peter. 2001. *Revolutionizing the Sciences*. Princeton: Princeton University Press. Contains an excellent brief discussion of Melancthon's involvement in science at Wittenberg.

Debus, Allen. 1978. *Man and Nature in the Renaissance*. Cambridge: Cambridge University Press. Includes an excellent elementary introduction to Hermeticism and Paracelsus.

De Santillana, Giorgio. 1955. *The Crime of Galileo*. Chicago: University of Chicago Press. Argues that a forged document entered into the inquisitorial records played a major role in the conflict between Galileo and the Catholic Church.

Descartes, René. 1955. *The Philosophical Works of Descartes*. 2 vols. Translated and edited by E. Haldane and G.R.T. Ross. New York: Dover Publications.

Desmond, Adrian. 1989. *The Politics of Evolution: Morphology, Medicine and Reform in Radical London*. Chicago: University of Chicago Press. Though not primarily focused on religion, this superb treatment of pre-Darwinian evolutionary ideas incorporates religious concerns as it explains why evolutionary ideas were more favored by political radicals than by conservatives in England.

———. 1997. *Huxley: From Devil's Disciple to Evolution's High Priest.* Boulder: Perseus Books. Excellent biography of the architect of Victorian agnosticism and scientific naturalism.

Duff, John William. 1979. Miracles in a World of Atoms? The Centrality of Providence in Walter Charleton's Mechanical Philosophy. Senior thesis, Harvard University.

Eamon, William. 1994. *Science and the Secrets of Nature: Books of Secrets in Medieval and Early Modern Culture.* Princeton: Princeton University Press. Includes a discussion of the roles of Christian humanists in promoting observational and experimental practices incorporated into the Book of Secrets tradition.

Efron, John. 1994. *Defenders of the Race: Jewish Doctors and Race Science in Fin-De Siécle Europe.* New Haven: Yale University Press. Argues that Jewish physicians played an important role in developing late nineteenth-century race theory and that they did so to promote ethnic pride in the face of anti-Semitism.

Eisen, Sydney, and Bernard Lightman. 1984. *Victorian Science and Religion: A Bibliography with Emphasis on Evolution, Belief, and Unbelief, comprised of works published from c. 1900–1975.* Hamden, Conn.: Archon Books. An unannotated listing of over 5,000 books and articles on the subject, broken down in a useful way by specific subject matter. Tremendously valuable, but demands that the user find some other method of assessing the quality of the listings.

Ellegård, Alvar. 1990. *Darwin and the General Reader: The Reception of Darwin's Theory of Evolution in the British Periodical Press, 1859–1872.* 1958. Reprint, Chicago: University of Chicago Press. The central chapters of this work, 5–8, focus on responses to Darwin in the religious press in England and constitute one of few careful empirical and quantitative studies of the responses to Darwin from various religious perspectives. It should be read by anyone interested in Darwin and religion. Chapter 9 also offers a worthwhile brief analysis of the responses to Darwin based on philosophy of science, which interacted with religion.

Epicurus. 1994. *The Epicurus Reader: Selected Writings and Testimonia.* Translated and edited by Brad Inwood and L. P. Gerson. Indianapolis: Hackett Publishing Company.

Farrington, Benjamin. 1953. *Greek Science.* Baltimore: Penguin Books.

Fauvel, John, Raymond Flood, Michael Shortland, and Robin Wilson, eds. 1988. *Let Newton Be!* Oxford: Oxford University Press. If you can only read one book about Isaac Newton, this should be it. Outstanding articles by experts on various aspects of Newton's life and works, with great illustrations.

Feingold, Mordechai, ed. 2002. *Jesuit Science and the Republic of Letters*. Cambridge: MIT Press. A broadly focused collection of articles presenting both a sense of the important scientific contributions of the Jesuits to early modern science and a sense of the constraints under which that science developed.

Feldhay, Rivka. 1987. "Knowledge and Salvation in Jesuit Culture." *Science in Context* 1:195–213. Argues that the practical orientation of Jesuit education allowed for a deviation from strict Thomist-Aristotelian logic and that both the Jesuits' attention to the possibility of free will and their interest in astronomy focused their attention on hypothetical objects.

Feldhay, Rivka, and Michael Heyd. 1989. "The Discourse of Pious Science." *Science in Context* 3:109–142. Fascinating attempt to characterize the differences between Jesuit and Calvinist science through the exploration of two specific teaching texts of the late seventeenth century.

Ferngren, Gary, ed. 2000. *The History of Science and Religion in the Western Tradition: An Encyclopedia*. New York and London: Garland Publishing. A useful one-volume reference work with 103 short articles by experts. Each article has a brief list of relevant secondary sources.

Feuerbach, Ludwig. 1957. *The Essence of Christianity*. Translated by George Eliot. 1854. Reprint, New York: Harper and Brothers.

———. 1969. *The Philosophy of Ludwig Feuerbach*. Edited by Eugene Kamenka. New York: Praeger Publishers.

Finocchiaro, Maurice. 1989. *The Galileo Affair: A Documentary History*. Berkeley and Los Angeles: University of California Press. Outstanding collection of primary sources relative to the relationship between Galileo and various factions within the Catholic Church.

Forbes, Geraldine Hancock. 1975. *Positivism in Bengal: A Case Study in the Transmission and Assimilation of an Ideology*. Calcutta: Minerva Associates. Includes the only discussion of the Religion of Humanity in India that I am aware of.

Force, James. 1985. *William Whiston: Honest Newtonian*. Cambridge: Cambridge University Press. Outstanding biography of the man who frequently expressed Newton's heterodox religious ideas in spite of the disastrous consequences to his own career and whose *New Theory of the Earth* was one of the most powerful early attempts to support the Genesis story of the creation with detailed scientific evidence.

French, Richard D. 1975. *Anti-vivisection and Medical Science in Victorian Society*. Princeton: Princeton University Press. A particularly good source for the religious views of Frances Cobbe, who was a leader of Victorian anti-vivisectionists.

Galileo Galilei. 1953. *Dialogue Concerning the Two Chief World Systems: Ptolemaic and Copernican*. Berkeley and Los Angeles: University of California Press.

Garin, Euginio. 1965. *Italian Humanism*. 1947. Reprint, New York: Harper and Row.

Giles of Rome. 1224. *On the Errors of the Philosophers*. Available at: http://medieval.ucdavis.edu/203/Giles.html.

Gillispie, Charles Coulston. 1959. *Genesis and Geology: The Impact of Scientific Discoveries upon Religious Beliefs in the Decades before Darwin*. New York: Harper and Row. The first major attempt to uncover the interactions between religious and scientific attitudes and ideas within geology leading up to the time of Darwin. Gillispie still shows strong traces of the positivist attitude, suggesting that geology in the early nineteenth century advanced only by throwing off its religious elements; but this is a good starting place for investigating the interactions between geology, natural theology, and religious literalism in the late eighteenth and early nineteenth century.

———. 1960. *The Edge of Objectivity: An Essay in the History of Scientific Ideas*. Princeton: Princeton University Press.

Glick, Thomas F. 1988. *The Comparative Reception of Darwinism*. Chicago: University of Chicago Press. Reprint of a volume based on a conference at the University of Texas in 1972 and first published in 1974. This edition contains an updated preface. Contains one of the few essays on the Islamic reception of Darwinism and an excellent essay by Harry Paul on Catholic responses.

Gonzales, Roberto. 2001. *Zapotec Science: Farming and Food in the Northern Sierra of Oaxaca*. Austin: University of Texas Press.

Gould, Stephen Jay. 1999. *Rocks of Ages: Science and Religion in the Fullness of Life*. New York: Ballantine Publishing Group. Well written defense of the position that science and religion cannot be in conflict because they have completely different aims and methods.

Greene, John C. 1959. *The Death of Adam: Evolution and Its Impact on Western Thought*. New York: New American Library. Outstanding exploration of the development of theories of temporal development in cosmology, geology, biology, and anthropology during the eighteenth and early nineteenth century. Includes many helpful diagrams and illustrations as well as extensive quotations from primary materials.

Gregory, Frederick. 1977. *Scientific Materialism in Nineteenth Century Germany*. Dordrecht: D. Reidel. The only reliable English language discussion of the social and religious context in which German materialism developed. Especially good on Ludwig Büchner, but includes a good discussion of Feuerbach.

———. 1992. *Nature Lost? Natural Science and the German Theological Traditions of the Nineteenth Century*. Cambridge: Harvard University Press. An excellent introduction to both academic and popular religious thought in Germany. Gregory makes extremely difficult material accessible.

Guthrie, W.K.C. 1965. *A History of Greek Philosophy, II*. Cambridge: Cambridge University Press.

Hakfoort, Casper. 1992. "Science Deified: Wilhelm Ostwald's Energeticist World-View and the History of Scientism." *Annals of Science* 49: 525–544. Deals extensively with Ostwald's role in promoting the scientistic religion of Monism.

Hall, A. R. and Marie Boas Hall, eds. 1962. *Unpublished Papers of Isaac Newton*. Cambridge: Cambridge University Press.

Hall, Marie Boas, ed. 1966. *Robert Boyle on Natural Philosophy*. Bloomington: Indiana University Press. A valuable single-volume collection of Boyle's writings that includes several whose major emphasis is religious.

———. 1970. *Nature and Nature's Laws*. New York: Harper and Row.

Hankins, Thomas. 1980. *Sir William Rowan Hamilton*. Baltimore: Johns Hopkins University Press.

Harris, Stephen J. 1989. "Transposing the Merton Thesis: Apostolic Spirituality and the Establishment of the Jesuit Scientific Tradition." *Science in Context* 3:29–65. Argues that many elements associated by Robert Merton with the Puritan ethos—a focus on diligence, practicality, learning as a form of worship, and so forth—also informed the ideology of the Jesuit order and encouraged Jesuit involvement in the sciences. Includes quantitative information about Jesuit contributions to various scientific fields, circa 1560–1780.

Harrison, Peter. 2001. *The Bible, Protestantism, and the Rise of Natural Science*. Cambridge: Cambridge University Press. Contends that Protestant changes in linguistic usages promoted the development of modern science.

Heilbron, John L. 1982. *Elements of Early Modern Physics*. Berkeley and Los Angeles: University of California Press. Chapter 2, "The physicists," contains an excellent analysis of Jesuit science from 1600 to 1789. Argues that Jesuits were especially important as experimental natural philosophers.

———. 1999. *The Sun in the Church: Cathedrals as Solar Observatories*. Cambridge: Harvard University Press. Detailed analysis of Catholic support for mathematical astronomy, circa 1540–1800. Argues that the Roman Catholic Church probably gave more financial and social support to astronomy than all other institutions combined over a long period of time and that the condemnation of Galileo had relatively little effect on Catholic mathematicians and astronomers. Demands some ability to follow geometrical arguments.

Helmstadter, Richard J., and Bernard Lightman, eds. 1990. *Victorian Faith in Crisis: Essays on Continuity and Change in Nineteenth Century Religious Belief*. Stanford: Stanford University Press. Eight of the eleven essays in this excellent collection focus on science and religion issues, though always with a social emphasis. Perhaps the best introduction to this field, though the importance of responses to evolutionary theory is not adequately explored.

Hermes Trismagistus. 1993. *Hermetica: The Ancient Greek and Latin Writings Which Contain the Religious or Philosophic Teachings Ascribed to Hermes Trismagistus*. Edited and translated by Sir Walter Scott. 1924–1926. Reprint, Boston: Shambhala Press.

Hippocrates. 1964. *The Theory and Practice of Medicine by Hippocrates*. Edited by Emerson C. Kelley. New York: Citadel.

Hobbes, Thomas. 1839. *The English Works of Thomas Hobbes*. Edited by Sir William Molesworth. London: John Bohn.

Holt, Niles R. 1971. "Ernst Haeckel's 'Monistic Religion.'" *Journal of the History of Ideas* 32:265–280. A good introduction to Monism. Should be supplemented with Hakfoort, 1992, for Ostwald's role.

Hovenkamp, Herbert. 1978. *Science and Religion in America: 1800–1860*. Philadelphia: University of Pennsylvania Press. Outstanding work. Complements Bozeman, 1977, by offering more contextual information and by treating New England Unitarians as well as the Old School Presbyterians associated with Princeton and the South. The later chapters take up key scientific topics, including polygenism, pre-Darwinian evolutionary theories, and geology, focusing on religious dimensions of American work in these areas.

Howell, Kenneth J. 2002. *God's Two Books: Copernican Cosmology and Biblical Interpretation in Early Modern Science*. Notre Dame: University of Notre Dame Press. A detailed and surprisingly readable account of one of the central foci of science and religion interactions during the sixteenth and seventeenth centuries. Howell argues that Protestant biblical hermeneutics interacted with Copernicanism in a wide variety of ways.

Huff, Toby. 1995. *The Rise of Early Modern Science: Islam, China, and the West*. Cambridge: Cambridge University Press. A study of why the scientific revolution occurred in the West rather than in Islam or in China, both of which were more scientifically and technologically advanced prior to the sixteenth century. One important emphasis is on the character of higher education in all three cultures and the specially positive attitude toward natural knowledge that emerged in the Christian context of European universities.

Hume, David. 1993. *Principle Writings on Religion, including Dialogues Concerning Natural Religion and The Natural History of Religion*. Oxford: Oxford University Press. *The Natural History of Religion* was first published in 1747, and the *Dialogs on Natural Religion* were only published in 1779, after Hume's death. Together they constitute one of the most powerful eighteenth-century skeptical attacks on religion in general and Christianity in particular.

Huxley, Thomas Henry. 1906. *Man's Place in Nature and Other Essays*. London: J. M. Dent and Sons.

Ignatius of Loyola. 1970. *The Constitutions of the Society of Jesus*. Edited by George Ganss. St. Louis: Institute for Jesuit Sources.

Irvine, William. 1955. *Apes, Angels, and Victorians: The Story of Darwin, Huxley, and Evolution*. New York: McGraw-Hill Book Co. Engagingly written but now somewhat dated discussion of the cultural context for and response to evolution by natural selection.

Jacob, James. 1972. "The Ideological Origins of Robert Boyle's Natural Philosophy." *Journal of European Studies* 2:1–21. One of the first articles to claim that Boyle's focus on natural religion preceded and shaped his science.

———. 1978. "Boyle's Atomism and the Restoration Assault on Pagan Naturalism." *Social Studies of Science* 8:211–233.

Jones, R. V. 1973. "James Clerk Maxwell at Aberdeen, 1856–1860." *Notes and Records of the Royal Society of London* 28:57–81.

Kant, Immanuel. 1902. *Prolegomena to Any Future Metaphysics*. Translated by Paul Carus. Lasalle, Ill.: Open Court.

———. 1965. *Immanuel Kant's Critique of Pure Reason*. Edited by Norman Kemp Smith. New York: St. Martin's Press.

Kelly, Alfred. 1981. *The Descent of Darwin: The Popularization of Darwinism in Germany, 1860–1914*. Chapel Hill: University of North Carolina Press. Chapter 5 contains an excellent short introduction to German religious responses to Darwin, including the Monism of Ernst Haeckel. For greater detail and sophistication, see Gregory, 1992.

Kepler, Johannes. 1859. *Opera*. Edited by Christian Frisch. Frankfurt.

Keynes, Randal. 2002. *Darwin, His Daughter, and Human Evolution*. New York: Riverhead Books. A charming biography of Darwin with a heavy emphasis on his interest in human evolution and the role of his children in promoting that interest. Very easy to read without being misleadingly simplistic.

Kohn, David, ed. 1985. *The Darwinian Heritage*. Princeton: Princeton University Press. Chapter 23 by Pietro Corsi and Paul Weindling is very valuable for its discussion of the reception of Darwin in German, French, and Italian contexts.

Kroll, Richard, Richard Ashcraft, and Perez Zagorin. 1992. *Philosophy, Science, and Religion in England, 1640–1700*. Cambridge: Cambridge University Press. An excellent set of papers focusing on the role of Latitudinarian Anglicanism in promoting science in seventeenth-century England.

Latour, Bruno. 1988. *Science in Action: How to Follow Scientists and Engineers Through Society*. Cambridge: Harvard University Press. The most notorious argument that all scientific change emerges out of conflict.

Lennon, Thomas M. 2000. "Cartesianism." Pp. 146–148 in Gary Ferngren, ed., *The History of Science and Religion in the Western Tradition: An Encyclopedia*. New York and London: Garland Publishing.

Lenoble, Robert. 1943. *Mersenne ou la naissance du mécanisme*. Paris: J. Vrin. The only useful book-length intellectual biography of Marin Mersenne, one of the most important Catholic advocates of the mechanical philosophy in the seventeenth century.

Lenzer, Gertrude, ed. 1975. *Auguste Comte and Positivism: The Essential Writings*. New York: Harper and Row. Contains extensive selections from Comte's religious writings.

Lightman, Bernard. 1987. *The Origins of Agnosticism: Victorian Unbelief and the Limits of Knowledge*. Baltimore: Johns Hopkins University Press. The best book on the subject. Contains biographically organized chapters on half a dozen Victorian agnostics, including T. H. Huxley.

Lindberg, David C., and Ronald L. Numbers, eds. 1986. *God and Nature: Historical Essays on the Encounter Between Christianity and Science*. Berkeley and Los Angeles: University of California Press. Outstanding collection of essays, several of which are separately listed. Covers the Patristic period into the twentieth century. One of the two best introductions (along with Brooke 1991) to the field of science and Christianity interactions.

———. 2003. *When Science and Christianity Meet*. Chicago: University of Chicago Press. Yet another excellent set of essays. Sequel to *God and Nature* but intended to reach a broader audience. The essay by Edward Larson on the Scopes trial as portrayed in "Inherit the Wind" is particularly good, as is Lindberg's essay on the Galileo affair.

Locke, John. 1958. *The Reasonableness of Christianity with a Discourse of Miracles*. Edited by I. T. Ramsey. Stanford: Stanford University Press.

———. 1959. *Essay Concerning Human Understanding*. Edited by A. C. Fraser. 1894. Reprint, New York: Dover.

Mackay, Donald. 1974. "'Complementarity' in Scientific and Theological Thinking." *Zygon* 9:225–244. Expression of the thesis that science and religion represent complementary, rather than conflicting, perspectives on the world.

Manuel, Frank. 1959. *The Eighteenth Century Confronts the Gods*. Cambridge: Harvard University Press. Excellent introduction to the eighteenth-century revival of classical interpretations of the origins of religion and the meaning of religious rituals and myths. Also includes one of the few readable accounts of the religious views of Giambattista Vico and Georg Hamann.

———. 1974. *The Religion of Isaac Newton*. Oxford: Oxford University Press. Still the best short account of Newton's personal religious beliefs in my view. Includes as appendices several of Newton's manuscripts providing rules for prophecy interpretation.

Mazlish, Bruce. 1998. *The Uncertain Sciences*. New Haven and London: Yale University Press.

McKeon, Richard, ed. 1947. *Introduction to Aristotle*. New York: The Modern Library. Traditionally, references to Aristotle are given in Becker numbers, which have been used since the mid-nineteenth century. However, McKeon was used primarily for citations within this book.

McKnight, Stephen A. 1989. *Sacralizing the Secular: The Rennaissance Origins of Modernity*. Baton Rouge: Louisiana State University Press. Among the best accounts of how Christian humanism promoted the development of applied sciences in the Renaissance.

———. ed. 1992. *Science, Pseudo-Science, and Utopianism in Early Modern Thought*. Columbia: University of Missouri Press. Excellent collection of articles on religion, the occult sciences, Hermeticism, and alchemy in the late Renaissance.

Merton, Robert K. 2002. *Science, Technology, and Society in Seventeenth Century England*. 1938. Reprint, New York: Howard Fertig. This classic initiated interest in the role of the Puritan ethos in promoting study of the natural sciences in early modern England. For responses see Cohen, 1990.

Miller, Perry. 1958. "Bentley and Newton." Pp. 271–278 in I. Bernard Cohen, ed., *Isaac Newton's Papers and Letters in Natural Philosophy*. Cambridge: Harvard University Press. While admitting the popularity of Bentley's Boyle Lectures, Miller is highly critical of Bentley.

Montgomery, John W. 1973. *Cross and Crucible: John Valentine Andreae (1586–1654)*. 2 vols. The Hague: Martinus Nijhoff. The most reliable discussion of this central figure in seventeenth-century Hermeticism and the institutionalization of early modern science. Volume 2 is a translation of *The Chymical Wedding of Christian Rosencreutz* which was taken (wrongly, Montgomery believes) to be the founding document of the Rosicrucian movement.

Moore, James R. 1979. *The Post-Darwinian Controversies: A Study of the Protestant Struggle to Come to Terms with Darwin in Great Britain and America, 1870–1900*. Cambridge: Cambridge University Press. The first chapter of the book contains one of the best critiques of the conflict thesis ever written. The remainder is a sustained and effective argument that evangelical Protestants had a much easier time accepting Darwin than more liberal Protestants.

———. 1986. "Geologists and Interpreters of Genesis in the Nineteenth Century." Pp. 322–350 in David C. Lindberg and Ronald L. Numbers, eds., *God and Nature: Historical Essays on the Encounter Between Christianity and Science*. Berkeley and Los Angeles: University of California Press. Outstanding short introduction to the interactions among scriptural geologists and secular geologists. Emphasizes American as well as British geologists.

Murphy, Howard. 1955. "The Ethical Revolt Against Christian Orthodoxy in Early Victorian England." *American Historical Review* 800ff. Outstanding article arguing that for many Victorian intellectuals a religious crisis preceded their turn toward science and secularism.

The New English Bible with the Apocrypha. 1970. London: Oxford University Press.

Newton, Isaac. 1733. *Observations Upon the Prophecies of Daniel and the Apocalypse of St. John*. London.

———. 1952. *Opticks*. 1730. Reprint, New York: Dover. The most widely available edition of the *Opticks*.

———. 1962. *Mathematical Principles of Natural Philosophy*. 2 vols. Edited by Florian Cajori. Berkeley and Los Angeles: University of California Press. The most widely available edition of the *Principia*, which was originally published in Latin in 1687.

Noll, Mark. 2000. "Evangelism and Fundamentalism." Pp. 298–306 in Gary Ferngren, ed., *The History of Science and Religion in the Western Tradition: An Encyclopedia*. New York: Garland Publishing.

Numbers, Ronald, and John Stenhouse, eds. 1999. *Disseminating Darwinism: The Role of Place, Race, Religion and Gender*. Cambridge: Cambridge University Press. Outstanding collection of articles, most of which address religious themes, extending consideration to regions and groups that have not been widely studied, including Ireland, Canada, New Zealand, Australia, American Catholics, American Jews, and American Blacks.

O'Connell, Marvin R. 1997. *Blaise Pascal: Reasons of the Heart*. Grand Rapids: Eerdmans. Very good account of Pascal's religious writings that explains the Jansenist movement that influenced him so greatly.

Odom, Herbert. 1966. "The Estrangement of Celestial Mechanics and Religion." *Journal of the History of Ideas* 27:533–548. Focuses on the Newtonian's use of the imperfection argument and the attack on that argument by Laplace.

O'Higgins, James. 1970. *Anthony Collins, The Man and His Works*. The Hague: Nijhoff. Excellent biography of the leader of eighteenth-century British Free Thought.

Olson, Richard. 1975. *Scottish Philosophy and British Physics, 1750–1880: A Study of the Foundations of the Victorian Scientific Style*. Princeton: Princeton University Press. Explores the relationships among religion, Scottish Common Sense philosophy, and natural philosophy in nineteenth-century Britain.

———. 1982. *Science Deified and Science Defied: The Historical Significance of Science in Western Culture*. Vol. 1, *From the Bronze Age to ca. 1620*. Berkeley and Los Angeles: University of California Press. A general cultural history of science with a significant emphasis on religion. This volume covers Hermeticism, Andreae, and Bacon in the Renaissance period.

———. 1983. "Tory-High Church Opposition to Science and Scientism in the Eighteenth Century: The Works of John Arbuthnot, Jonathan Swift, and Samuel Johnson." Pp. 171–204 in John G. Burke, ed., *The Uses of Science in the Age of Newton*. Berkeley and Los Angeles: University of California Press. Counters arguments by Margaret Jacob that High Church Anglicans were particularly antagonistic to Newtonian science by showing that many were hostile to all varieties of natural theology and scientific practice.

———. 1987. "On the Nature of God's Existence, Wisdom, and Power: The Interplay Between Organic and Mechanistic Imagery in Anglican Natural Theology—1640–1740." Pp. 1–48 in Frederick Burwick, ed., *Approaches to Organic Form*. Dordrecht and Boston: D. Reidel Publishing Company. Explores the tension within Anglican natural theology between mechanical philosophies and those that insisted upon the uniqueness and self-moving characteristics of living beings. Argues that there was an important place within Anglicanism for modified versions of neo-Platonist ideas.

———. 1990. *Science Deified and Science Defied: The Historical Significance of Science in Western Culture.* Vol. 2, *From the Early Modern Age through the Early Romantic Era, ca. 1640 to ca. 1820.* Berkeley and Los Angeles: University of California Press. A general cultural history of science with significant emphasis on religion. The chapters on Newton and on the Romantic reaction are likely to be of greatest interest.

O'Malley, John, et al., eds. 1999. *The Jesuits: Cultures, Sciences, and the Arts, 1540–1773.* Toronto: University of Toronto Press. Nearly a third of the thirty-two articles in this collection discuss Jesuit sciences before the suppression of the order by Clement XIV. They provide an excellent introduction that should be supplemented by Heilbron, 1999, on Jesuit astronomy.

Osler, Margaret. 1994. *Divine Will and the Mechanical Philosophy: Gassendi and Descartes on Contingency and Necessity in the Created World.* Cambridge: Cambridge University Press. An outstanding investigation of the ways in which theological considerations, especially regarding God's absolute power, shaped the formation of the two most important traditions of mechanical philosophy in the seventeenth century. Presents subtle arguments in an accessible way.

———, ed. 2000. *Rethinking the Scientific Revolution*. Cambridge: Cambridge University Press. This work illustrates the extent to which theological issues have penetrated recent work on the scientific revolution. Of the fifteen papers, twelve deal with science and religion interactions.

Ovitt, George. 1986. *The Restoration of Perfection*. New Brunswick, N.J.: Rutgers University Press. An excellent introduction to medieval and early modern millenarianism.

Paracelsus. 1951. *Paracelsus: Selected Writings*. Edited by Jolande Jacobi. Princeton: Princeton University Press.

———. 1999. *Paracelsus: Essential Readings*. Edited and translated by Nicholas Goodrick-Clarke. Berkeley, Calif.: North Atlantic Books.

Paul, Harry. 1979. *The Edge of Contingency: French Catholic Reaction to Scientific Change from Darwin to Duhem*. Gainesville: University Press of Florida. The most reliable book-length coverage of French Catholic relations with science.

Petrarch, Francesco. 1910. *Petrarch's Letters to Classical Authors*. Translated by M. E. Cosemya. Chicago: University of Chicago Press.

Ray, John. 1705. *The Wisdom of God, Manifested in the Works of Creation*. 5th ed. London.

Redondi, Pietro. 1987. *Galileo, Heretic*. Chicago: University of Chicago Press. Controversial work suggesting that it was Galileo's enthusiasm for the atomic philosophy of Lucretius rather than his support of Copernican astronomy that led to his problems with the Catholic Inquisition.

Reid, Thomas. 1895. *The Philosophical Works of Thomas Reid*. Edited by Sir William Hamilton. Edinburgh: James Thin.

Roberts, Jon. 1988. *Darwinism and the Divine in America: Protestant Intellectuals and Organic Evolution, 1859–1900*. Madison: University of Wisconsin Press. Best overview of this topic. Because he uses a wide range of sources, largely drawn from quarterly magazines, he avoids overemphasis on a few figures.

Royle, Edward. 1971. *Radical Politics 1790–1900: Religion and Unbelief*. London: Longmans. A good introduction to scientistic working class religious movements in the early nineteenth century.

———. 1974. *Victorian Infidels: The Origin of the British Secularist Movement, 1791–1866*. Manchester: Manchester University Press. A detailed investigation of the most successful working-class anticlerical movement in Britain.

Ruderman, David. 1995. *Jewish Thought and Scientific Discovery in Early Modern Europe*. Detroit: Wayne State University Press. Very accessible introduction to the interactions between Jewish culture and scientific culture from the eleventh century through the eighteenth.

Schmitt, Charles B. 1983. *Aristotle and the Renaissance*. Cambridge: Harvard University Press. My choice for the best introduction to the combination of continuing scholastic and renewed humanistic approaches to Aristotelian philosophy in the early modern period. Though not explicitly focused on religious issues, it provides important background information.

Schofield, Robert. 1963. "Joseph Priestley: Theology, Physics, and Metaphysics." *Enlightenment and Dissent* 2:69–81.

Secord, James. 2000. *Victorian Sensation: The Extraordinary Publication, Reception, and Secret Authorship of* Vestiges of the Natural History of Creation. Chicago: University of Chicago Press. The most detailed study available of the reaction to pre-Darwinian evolutionary ideas, including reactions from various religious perspectives. Long, but filled with amazing detail.

72 Nobel Laureates, 17 State Academies of Science and Seven Other Scientific Organizations. 1986. *Amicus Curiae*. Brief in support of Appelles Don Aguilard et al. v. Edwin Edwards in his official capacity as Governor of Louisiana et al. Evolutionary scientists' response to the Louisiana attempt to mandate equal time for creation science. Interesting in part because it articulated a definition of science that was used by the Supreme Court to exclude creation science.

Shapin, Steven, and Simon Shaffer. 1985. *Leviathan and the Air Pump: Hobbes, Boyle, and the Experimental Life*. Princeton: Princeton University Press. Difficult reading, but an outstanding discussion of the religious and political dimensions of the conflict between experimental and rationalist versions of the mechanical philosophy. Especially valuable for the treatment of Hobbes as a scientist.

Shapiro, Barbara. 1968. "Latitudinarianism and Science in Seventeenth-Century England." *Past and Present* 40:16–41. First major article arguing that Latitudinarians rather than Puritans were the major supporters of science in seventeenth-century England. Superseded by writings of James and Margaret Jacob.

———. 1969. *John Wilkins: 1614–1672: An Intellectual Biography*. Berkeley and Los Angeles: University of California Press. A useful biography of one of the key figures among the founders of the Royal Society. Offers a valuable discussion of the problems associated with identifying the religious loyalties of many seventeenth-century intellectuals.

———. 1983. *Probability and Certainty in Seventeenth-Century England: A Study of the Relationships between Natural Science, Religion, History, Law, and Literature*. Princeton: Princeton University Press. The best single introduction to probabilism in philosophy, religion, and legal thought in seventeenth-century England.

———. 1991. "Early Modern Intellectual Life: Humanism, Religion, and Science in Seventeenth-Century England." *History of Science* 29:45–71. A very insightful overview of the connections among Christian humanism, Latitudinarian religion, and natural philosophy.

Shaw, George Bernard. 1921. *Back to Methuselah: A Metabiological Pentateuch.* New York: Brentano's.

Smith, Crosbie. 1999. *The Science of Energy: A Cultural History of Energy Physics in Victorian Britain.* Chicago: University of Chicago Press. Though not primarily aimed at discussing science and religion interactions, this book nonetheless does an excellent job of explaining what religious considerations made the approaches to energy of a group of "North British" physicists differ from that of John Tyndall, for example.

Sprat, Thomas. 1702. *History of the Royal Society.* London.

Stanford, W. B. 1980. *Enemies of Poetry.* London: Routledge and Kegan Paul.

Stocking, George W. 1968. *Race, Culture, and Evolution.* New York: Free Press. Major treatment of Victorian anthropology, one of whose major emphases was the investigation of the evolution of religion.

Sullivan, Robert. 1982. *John Toland and the Deist Controversy.* Cambridge: Harvard University Press. Outstanding introduction to the most notorious pantheist in early eighteenth-century Britain.

Swetlitz, Marc. 1999. "American Jewish Responses to Darwin and Evolutionary Theory, 1860–1890." Pp. 209–245 in Ronald Numbers and John Stenhouse, eds., *Disseminating Darwinism: The Role of Place, Race, Religion, and Gender.* Cambridge: Cambridge University Press. Argues that Reform Jewish intellectuals used evolutionary ideas to support their split from traditionalists after a period of generally negative responses to Darwin among American Jews.

Tertullian. 1896–1903. "Ad Nationes." In Peter Holmes, trans., *The Anti-Nicene Fathers.* Edited by Alexander Roberts and James Donaldson. Volume 3. New York: Charles Scribner's Sons.

Topham, Jonathan. 1992. "Science and Popular Education in the 1830's: The Role of the *Bridgewater Treatises.*" *British Journal for the History of Science* 25. Excellent introduction to the role of the Bridgewater Treatises in promoting natural theology across class lines.

Toulmin, Stephen, and June Goodfield. 1977. *The Discovery of Time.* 1965. Reprint, Chicago and London: University of Chicago Press. Extremely wide-ranging and readable exploration of the idea that both nature and human society change over time, beginning with pre-Greek mythologies and moving through the nineteenth century. Relates these ideas to the scriptural story of creation.

Toumey, Christopher. 1994. *God's Own Scientists: Creationists in a Secular World*. New Brunswick, N.J.: Rutgers University Press.

Turner, Frank M. 1978. "The Victorian Conflict between Science and Religion: A Professional Dimension." *Isis*, 69:356–376. The classic statement that professional jealousies played a central role in the Victorian conflicts between professionalizing scientists and clergy.

Tyndall, John. 2000. "Address Delivered before the British Association for the Advancement of Science Assembled at Belfast, with Additions." Pp. 359–385 in A.S. Weber, ed., *19th Century Science: An Anthology*. Peterborough, Canada: Broadview Press.

Walker, D.P. 1981. *Unclean Spirits: Possession and Exorcism in France and England in the Late Sixteenth and Early Seventeenth Centuries*. Philadelphia: University of Pennsylvania Press.

Webster, Charles. 1976. *The Great Instauration; Science, Medicine, and Reform 1626–1660*. New York: Holmes & Meier. Massive and detailed study of science among millenarian radical Protestants before and during the English Revolution. Special emphasis is on Baconian experimental science and the circle surrounding Samuel Hartlib.

———. 1982. *From Paracelsus to Newton: Magic and the Making of Modern Science*. Cambridge: Cambridge University Press. Excellent short work focusing on the role of Paracelsan ideas, especially those associated with prophecy interpretation, in seventeenth-century English intellectual life.

Westfall, Richard S. 1958. *Science and Religion in Seventeenth-Century England*. New Haven: Yale University Press. One of the first works to emphasize the extent to which many major English scientists turned to writing natural theological works. Excellent in characterizing the content of many of these works, but more inclined than later scholars to see natural theology as an attempt to reconcile presumptive conflicts between the mechanical philosophy and traditional religious doctrines.

Westman, Robert, and J.E. McGuire. 1977. *Hermeticism and the Scientific Revolution*. Berkeley and Los Angeles: University of California Press. One of the strongest attacks on the arguments of Frances Yates. The focus on heliocentrism is particularly valuable.

Whiston, William. 1708. *A New Theory of the Earth*. 2d ed. London.

———. 1725. *A Supplement to the Literal Accomplishment of Scripture Prophecies.* London.

White, Andrew Dickson. 1965. *A History of the Warfare of Science with Theology in Christendom.* New York: The Free Press. Abridgment of the original 1896 classic statement of the conflict thesis.

Wilde, C. B. 1980. "Hutchinsonianism, Natural Philosophy, and Religious Controversy in Eighteenth-Century Britain." *History of Science* 18:1–24. Excellent introduction to the character and context of the scripturally based Hutchinsonian alternative to Newtonian natural philosophy.

Wilson, David Sloan. 2002. *Darwin's Cathedral: Evolution, Religion, and the Nature of Society.* Chicago: University of Chicago Press. A fascinating modern extension of Darwin's account of the importance of religion for human social evolution from *The Descent of Man.* Argues in addition that religions function as adaptive units and applies evolutionary ideas to the rise of Calvinism in particular.

Wright. T. R. 1986. *The Religion of Humanity: The Impact of Comtean Positivism on Victorian Britain.* Cambridge: Cambridge University Press. The best account of positivist religion in England.

Yates, Frances A. 1964. *Giordano Bruno and the Hermetic Tradition.* New York: Random House. This work, more than any other, stimulated interest in Hermeticism among historians of science. Though it has drawn severe criticism regarding certain details, it initiated a rich tradition of scholarship.

Yolton, John. 1983. *Thinking Matter: Materialism in Eighteenth-Century Britain.* Minneapolis: University of Minnesota Press. The best introduction to the interaction between materialist natural philosophies and Christianity in the eighteenth century.

Zammito, John H. 2002. *Kant, Herder, and the Birth of Anthropology.* Chicago: University of Chicago Press.

Index

About the Author

RICHARD G. OLSON is Professor of History and Willard W. Keith Fellow in the Humanities at Harvey Mudd College. His work has focused on the interrelationships between the natural sciences and other cultural domains, including moral philosophy, the social sciences, political ideology, and religion. His publications include *Science Deified and Science Defied* (vol. 1, 1982; vol. 2, 1990) and *The Emergence of the Social Sciences, 1642–1792* (1993).